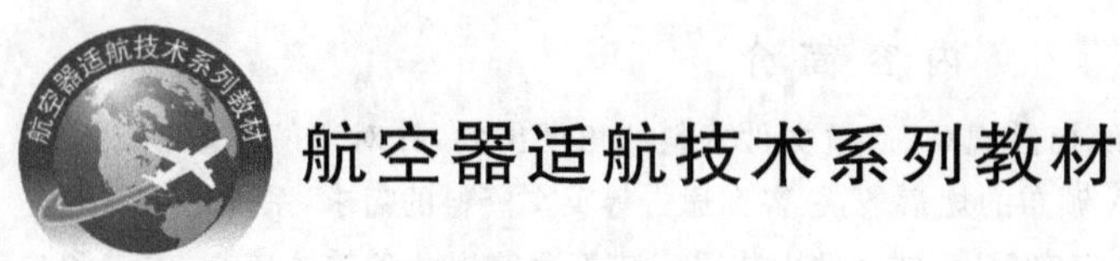

适航英语

Airworthiness English

苏 艳 ［乌克兰］Oleksiy Chernykh 顾 新 张迎春 编

北京航空航天大学出版社

内容简介

本书是从理解整个适航领域的相关规章、适航文件资料的角度出发，针对适航专业领域学生和从业技术人员理解 FAA、EASA 颁布的规章条款、咨询通告等英文资料的需求，系统地诠释航空器性能适航、结构适航、系统适航、审定程序、辅助动力装置安装等适航规章条款的核心、关键和难点词汇及其应用。

本书可用作适航专业英语教材，也可为航空领域从业人员更好地理解国际适航规章条款、适航文件资料提供参考，还可作为适航专业或相关专业的研究生、技术人员学习适航专业英语术语的参考书。

图书在版编目(CIP)数据

适航英语/苏艳等编. -- 北京 ：北京航空航天大学出版社，2021.2

ISBN 978-7-5124-3445-5

Ⅰ. ①适… Ⅱ. ①苏… Ⅲ. ①飞机—适航性—英语 Ⅳ. ①V221

中国版本图书馆 CIP 数据核字(2021)第 024657 号

适航英语

Airworthiness English

苏 艳 ［乌克兰］Oleksiy Chernykh 顾 新 张迎春 编

策划编辑 董 瑞 责任编辑 江小珍

*

北京航空航天大学出版社出版发行

北京市海淀区学院路 37 号(邮编 100191) http://www.buaapress.com.cn

发行部电话：(010)82317024 传真：(010)82328026

读者信箱：goodtextbook@126.com 邮购电话：(010)82316936

北京建宏印刷有限公司印装 各地书店经销

*

开本：710×1 000 1/16 印张：13.25 字数：290 千字

2021 年 8 月第 1 版 2021 年 8 月第 1 次印刷 印数：1 000 册

ISBN 978-7-5124-3445-5 定价：46.00 元

前　言

如今航空业是国际性联系较强的产业，需要一大批国际化专业人才。本书站在理解国际适航相关规章的角度，系统地诠释适航规章条款的核心、关键和难点词汇及其应用，以帮助相关专业学生和从业人员更好地理解 FAA、EASA 颁布的适航规章、咨询通告等英文资料。适航专业术语源自国际适航规章，无法由个人撰写相应的规章条款，因此，本书英文原文内容源自国际适航规章。

全书共有 7 章，包括航空器性能适航、结构适航、系统适航、审定程序、辅助动力装置安装等涵盖核心、关键或难点术语的适航规章条款，以及适航组织机构和国家的责任等内容。每章都列出相应规章的核心、关键和难点词汇的中英文，并对疑难句子进行专门释义。为帮助读者掌握专业术语、理解规章条款内容，每章最后还配有相关习题。

本书可以作为适航专业英语教材，也可为航空领域从业人员更好地理解国际适航规章条款、适航文件资料提供参考，还可作为适航专业或相关专业的研究生、技术人员学习适航专业英语术语的参考书。

由于作者水平有限和编写时间仓促，书中可能存在许多不足或错误之处，恳请读者指正。

本书的出版得到了南京航空航天大学左洪福教授、孙有朝教授、陈果教授等的关心与支持，课题组的梁学瑞、邢翔宇、Shikder Shovon 三位研究生在录入与翻译、校核方面也给予了帮助，在此一并致以诚挚的谢意。

编　者

2020 年 11 月

Contents

Chapter 1 Airworthiness Organizational Structure and States' Responsibilities

Part I Regulatory Document Text Example

[from DOC9760 ICAO[①] Airworthiness Manual Part 2]

1.1 State Airworthiness Responsibilities

1.1.1 Obligations Under the Convention on International Civil Aviation

The *Convention on International Civil Aviation* was signed in Chicago on 7 December 1944. The preamble states in part, "Whereas the future development of international civil aviation can greatly help to create and preserve friendship and understanding among the nations and peoples of the world. Therefore undersigned governments have agreed on certain principles and arrangements in order that international civil aviation is developed in a safe and orderly manner".

With respect to airworthiness, Article 29 of the Convention provides that every aircraft of a Contracting State, engaged in international navigation, carries a Certificate of Registration and a Certificate of Airworthiness. Article 31 of the Convention provides that the Certificate of Airworthiness is issued or rendered valid by the State in which the aircraft is registered. Furthermore, Article 33 of the Convention requires that Contracting States recognize as valid the Certificate of Airworthiness issued or rendered valid by the State of Registry provided that the requirements under which the certificate was issued or rendered valid are equal to or above the minimum standards established by ICAO. Article 54 of the Convention allows ICAO to issue Standards and Recommended Practices (SARPs) and to designate them as Annexes to the Convention. The Annexes largely covering the airworthiness of aircraft are: Annex 6—Operations of Aircraft; Annex 8—Airworthiness of Aircraft.

(Note: Annex 1—Personnel Licensing, Annex 7—Aircraft Nationality and Registration Marks and Annex 16—Environmental Protection, also provide the

---

① ICAO is the abbreviation of International Civil Aviation Organization.

requirements for the licensing of maintenance personnel, registration of aircraft and noise certification, respectively.)

Annex 8 outlines a framework of airworthiness standards for the design and manufacture of large aeroplanes, helicopters, small aeroplanes, engines and propellers. Annex 6 outlines minimum requirements for the operation and maintenance of aircraft. Annex 6 is issued in three Parts: Part Ⅰ—International Commercial Transport—Aeroplanes, Part Ⅱ—International General Aviation—Aeroplanes, and Part Ⅲ—International Operations—Helicopters.

Article 12 of the Convention points out that each Contracting State undertakes to keep its own regulations in these respects uniform, to the greatest extent possible, with those established from time to time under this Convention. Through national regulations, States are expected to implement and enforce the Standards contained in the Annexes to the Convention.

Article 37 of the Convention requires each Contracting State to collaborate in securing the highest practical degree of uniformity in regulations, standards, procedures and organization in relation to aircraft. Contracting States have the responsibility for the safe operation and performance of maintenance in accordance with the SARPs. Each State should develop its own airworthiness standards based on the framework provided in Annex 8, or adopt those already developed by another State.

In the development of national airworthiness regulations and rules, the State of Registry is responsible to make certain that every aircraft on its register conforms to the approved type design in accordance with the airworthiness code it has adopted or accepted for that class of aircraft. The State of Registry also has the responsibility to make certain that every aircraft on its register is maintained in an airworthy condition throughout its operational service life. Therefore, effective continuing airworthiness requirements are most important. Although methods of discharging the foregoing State airworthiness responsibilities may vary, and in some cases may involve the transfer of certain tasks to authorized organizations or other States, such arrangements do not relieve the State of Registry from its overall responsibility.

The State of Registry is responsible for the maintenance performed by an approved maintenance organization (AMO) or under an acceptable equivalent system. A review of the aircraft register and of the authorizations granted to an AMO or under an acceptable equivalent system will assist the State in determining if the operations are within its safety oversight capability. Where this is not the case, the State is urged to take immediate and appropriate actions to enhance the

State oversight system.

1.1.2 Discharge of State Responsibilities

The airworthiness regulatory system to be established will vary depending upon the level of complexity and scope of aviation activity within the State. The airworthiness regulatory system should include:

(a) Drafting and amendment of rules relating to the airworthiness of aircraft;

(b) Issuance, acceptance or validation of type certificate of aircraft, engine and propeller;

(c) Approval and continued inspection of approved design and production organization of aircraft and parts;

(d) Registration of aircraft;

(e) Certification and approval of initial applications of air operators (airworthiness aspects);

(f) Continued inspection and surveillance of certificated air operators (airworthiness aspects);

(g) Issuance, acceptance or validation of noise certificate;

(h) Issuance, renewal and continuing validation of the Certificate of Airworthiness;

(i) Approval of maintenance programme;

(j) Approval of modifications and mandatory inspection;

(k) Approval of repairs;

(l) Approval and continuing inspection of AMOs;

(m) Monitoring and control of mandatory continuing airworthiness information (MCAI);

(n) Approval and continued inspection of approved maintenance training organizations; and

(o) Licensing of aviation personnel.

Through the proper process of registering aircraft and the issuance of the certificates and approvals, the State ensures public safety and interest are addressed. Furthermore, the State will be able to exercise appropriate influence and control of airworthiness activities without encroaching upon its air operators, design organizations, manufacturers and maintenance organizations' responsibilities for safety.

1.2 Primary Aviation Legislation

Primary aviation legislation should contain provisions to establish a civil aviation administration that proactively supervises and regulates civil aviation activities. The legislation should contain provisions for the approval and safety oversight of air operators, aircraft maintenance organizations, and maintenance training organizations for the safe operation of aircraft. It should also establish requirements for the qualification and licensing of maintenance personnel. The legislation should also reference appropriate support functions with the establishment of an airworthiness inspection division (AID) and airworthiness engineering division (AED). The AID and AED duties should include the registering of aircraft, airworthiness certification of air operators, issuance and renewal of the Certificate of Airworthiness, issuance of documents attesting noise certification, approval of maintenance organization, approval of maintenance programme, the evaluation and issuance of modification and repair approvals, and the evaluation and issuance of design approvals and approval of manufacturing organizations. In addition to the certification duties it will be necessary for the legislation to stipulate the need for the State to perform surveillance of its aviation industry adequate to ensure that at least annual audits are conducted of its certified air operators, AMOs, training organizations, and design and production organizations. The surveillance programme should sufficiently cover the complexity of the State's aviation industry. The legislation should also contain provisions for the issuance and compliance of mandatory continuing airworthiness information (MCAI), as necessary.

With the enactment of primary aviation legislation the State can develop and promulgate civil aviation regulations and practices, including airworthiness regulations consistent with the Annexes. In the development of these regulations, the State is responsible to ensure that the regulations are consistent with and relevant to the Convention, Annexes and its aviation industry. Further, the State will need to ensure a process exists for the amendment of its regulations and the notification to ICAO of differences, when necessary.

1.3 State Regulatory System

1.3.1 General

The State's law and regulations should conform to the Chicago Convention and its Annexes. The regulations may be viewed as minimum requirements that may be exceeded by the certificate holder. Contracting States are responsible for developing regulations with sufficient detail that, at a minimum, support the complexity of the aviation industry within their States and achieve a satisfactory level of safety. The development of regulations should include an amendment process that includes publishing the amendment for comment within the State. The comment review process should consist of a review, consideration, disposition, and appropriate revision of any proposed amendment. The review should take into consideration the benefit to aviation safety, the public's ability to comply with the amendment and any potential undue burden the amendment may pose on the industry.

Annexes are published to supplement the Convention, as referenced in Article 54. Annexes specify SARPs that a Contracting State can reference in order to ensure the highest practicable degree of uniformity in its national regulations. Regulations developed by States should be supported by law. The law should address the types of regulations necessary to facilitate a capable authority oversight system and licensing system. The regulations should also address the requirements needed for the applicability, qualification, certification, design, manufacturers, maintenance organizations, training organizations, air operators, and aircraft engaged in international air transportation to be consistent with the Annexes. In some instances States may have a need to develop regulations that go beyond the minimum standards. In this case, States should expect the certificate holder, e. g. air operator or AMO, to take responsibility in providing a specific expectation, for example, approval of major modification and repair designs by the CAA①, prior to returning the aircraft to service.

A State's regulations should be worded in legal terms and written so they can be used by the authority in licensing, certificating and approvals during the day-to-day functions of the State safety oversight system. The regulations should be performance based and form a framework of a minimum standard unless the provision calls for a higher standard.

The State has an option of adopting other Contracting States' regulations. Adopting another State's regulations has some advantages but should be considered only after ensuring the other State's regulations are consistent with the ICAO Standards. The complexity of the other State's aviation environment should also be considered. A State with a limited aviation environment should be careful not to place an undue burden on its aviation community by adopting complex regulations. A State may consider adapting its regulations to meet the needs of its aviation environment while still maintaining consistency with other States. A State adopting regulations from another Contracting State would be expected to maintain a currency with the amendments of the adopted regulations and report differences to ICAO in accordance with Article 38 of the Convention.

Article 38 of the Convention specifies if the State finds it impracticable to

① CAA is the abbreviation of Civil Aviation Authority.

comply in all respects with any international standards or to bring its own regulations in full accord with international standards as amended, or if it deems it necessary to adopt regulations or practices differing from ICAO Standards, the State immediately notifies ICAO of the differences between its own practice and that established by the international standards. ICAO is obliged to immediately notify all other States of the difference which exists between one or more aspects of the Standard and the corresponding national practice of the notifying State. Several Articles of the Convention make it clear that if standards adopted by a State are lower than those required by ICAO, aircraft, certificate holders, or persons with licenses or certificates may not be able to participate in international air navigation, except with the permission of the State or States whose territory is entered. A State may request blanket permission on behalf of its certificate holders or licence holders.

Compliance with the regulations is not optional. Occasionally, there might be an instance where there is a geographical, physical or operational environmental problem that places an undue burden upon the certificate holder in complying with the relevant regulation. With well documented regulations in place a State may grant an exception or exemption to a regulation provided there exists a mechanism within the State's criteria to establish that exception or exemption. The relief granted should be considered the exception and not the norm. The person requesting relief should prove that the current requirement places an undue burden upon the certificate holder. A risk assessment should also be conducted to ensure that the exemption sought will not cause an adverse impact towards safety. The request for exemption should also contain an equivalent level of safety commensurate with the regulation and will be practiced should the exemption be granted by the State. If an exemption is granted it should be accompanied by conditions and limitations, including a time limitation to the exemption. The authority granting the exemptions should monitor such exemptions to ensure that continuation of the relief is warranted.

1.3.2 Airworthiness Regulations

The airworthiness regulations developed, adopted or accepted by the State should include provisions for:

(a) mandatory registration of all aircraft;

(b) implementation of the airworthiness provisions meeting the requirements of the Convention and the Annexes;

(c) all aircraft on the State's aircraft register to meet relevant airworthiness criteria approved or adopted by the State;

(d) the issuance, validation or acceptance of the type certificate for aircraft intended to be entered on the State's aircraft register;

(e) the issuance of production certificates or approvals of manufacturing organizations when applicable;

(f) the issuance, renewal, validation or acceptance of aircraft certificates of airworthiness;

(g) the issuance of export certificates of airworthiness;

(h) the issuance or acceptance of ADs[①], bulletins and orders;

(i) the issuance, amendment, cancellation and suspension of airworthiness approvals, licences and certificates;

(j) the authorization of persons or organizations, on behalf of the CAA, to perform particular tasks in relation to the design, manufacture and maintenance of aircraft, components and parts for the issuance of State approvals, licences and certificates, as appropriate;

(k) the authorization of persons and organizations, on behalf of the CAA, to inspect and test aircraft, aircraft components, standard parts, materials or processes and systems for the purpose of ascertaining whether the processes and activities covered by an approval, licence or certificate have been carried out in a satisfactory manner, as appropriate.

1.4 Airworthiness Organization

1.4.1 Structure of the CAA

Pursuant to delegated authority, the Director General of Civil Aviation (DGCA) should establish an effective organization and employ the necessary qualified personnel to carry out its responsible functions. The structure and size of the CAA's airworthiness organization will vary considerably depending on the number, size and complexity of civil air operations (commercial and private) in the State and on the size and scope of the State's aviation manufacturing and maintenance industry.

In deciding upon the required airworthiness organizational structure, the DGCA should review the requirements for certification and surveillance of air operators as outlined in this Airworthiness Manual and also in the Manual of Procedures for Operations Inspection, Certification and Continued Surveillance (Doc 8335), in light of the number and size of potential air operators in the State. The DGCA should consider the level of civil aviation activity and the size of the

---

① ADs is the abbreviation of Airworthiness Directions.

State's aviation manufacturing and maintenance industry when establishing the organizational structure. In those States where there are extensive aviation operations, manufacturing and maintenance, it will generally be necessary to establish within the CAA airworthiness organization an airworthiness engineering division (AED) and an airworthiness inspection division (AID). The establishment and functioning of these two divisions are discussed in this manual. To be effective, the CAA should provide an adequate level of administrative support, including comprehensive information technologies, facilities, and means of transportation for members of the organization.

The Convention requires that a number of approvals be issued by the State, but the CAA may wish to authorize an organization or a delegated person to make approvals on behalf of the CAA. A system of delegations implemented by the CAA will generally satisfy this requirement, providing it incorporates the following features:

(a) The national law and/or regulation permit the CAA to designate its functions.

(b) The standards to be achieved are clearly documented by the CAA. A designee can apply only a documented standard approved by the CAA.

(c) The designees are required to meet technical and regulatory competency requirements and are authorized to make approvals only in areas of their demonstrated competence.

(d) The CAA has an interest in the continuing proficiency of the designees and monitors their continued training so that they remain competent in the fields of their authorization.

(e) The designees' procedures have been approved by the CAA, and the CAA audits the designees to ensure they follow those procedures. The procedures should clearly identify where an approval is made, and will normally include a clear differentiation between the development of data and the approval of that data.

(f) The basis for making the approval is clearly documented.

(g) The designees make approvals for and on behalf of the CAA.

An approval made by a correctly authorized designee will be accepted as if it was made by the CAA. However, responsibility for the issuance of the approval remains with the State.

[Note: The Safety Oversight Manual (Doc 9734)—Part B (The Establishment and Management of a Regional Safety Oversight System) provides guidance on such arrangements.]

1.4.2 Staffing and Training

1. Qualification of AID Inspectors by Academic Study

Ideally, an AID inspector should be as qualified as the personnel to be inspected or supervised. This is usually accomplished by the inspectors having previous engineering or aircraft maintenance experience. There may be occasions where there is a shortage of such qualified AID inspectors. As an alternative and on a case-by-case basis, AID inspectors may be deemed qualified by the successful completion of a course of relevant academic aeronautical study at a recognized approved training organization or university.

To perform their duties adequately it is important that these new inspectors undergo a comprehensive technical on-the-job training programme that provides the inspector with the necessary expertise, experience and skills necessary to perform the tasks required of an AID inspector. The new inspector should be teamed with an experienced inspector who will ensure the on-the-job training is performed and documented. AID inspectors holding academic credentials with no previous aircraft maintenance experience should only be appointed in extraordinary circumstances.

2. Training Requirements

The CAA should determine the minimum qualifications for their technical personnel performing safety oversight functions and also provide for their technical training on an initial and recurrent basis. Additionally, periodic practical and specialized technical training including supervisory courses will enable the technical personnel to perform their duties effectively. Training should not be limited to strictly professional elements; technical personnel should receive training on subjects such as applicable CAA regulations, inspector handbooks, auditing techniques, safety management systems (SMS) and quality systems, Human Factors principles, enforcement procedures and topics dealing with advances in aviation technology.

The CAA should have a structured programme to educate the personnel on new CAA requirements, policies and procedures as they are being implemented. To keep personnel abreast of new industry developments a training programme should be developed that provides at regular intervals (initial and recurrent) technical training to gain first-hand knowledge of new developments, including management principles. As a general policy, it is not desirable for CAA personnel to obtain technical qualifications from those entities under their direct regulatory jurisdiction.

1.4.3 Airworthiness Engineering Division (AED) Responsibilites

States with a significant aviation manufacturing industry should establish an

AED within the CAA airworthiness organization. The size and structure of the AED should be appropriate to the aviation manufacturing industry and the various types of aircraft on the State's aircraft register. The AED's activity will normally be directed toward design approvals, type certification, manufacturing approvals, evaluations of modifications and repairs proposed by manufacturers, aircraft owners, air operators, and AMOs, to correct deficiencies of aircraft already in service.

In States with significant manufacturing activity it may be useful to organize the AED along functional lines by organizing sections dealing with specific technical specialities, such as structures, propulsion, electrical and mechanical systems (including software), certification of production organizations, and surveillance and oversight functions. Furthermore, specialized areas such as non-destructive inspection, plating, welding, software quality assurance and special manufacturing processes should be considered.

When the physical size of the State is large and the level of aviation activity is relatively high, it may be necessary to establish regional offices in the proximity of the aviation industry. In such cases, it is necessary that proper lines of communication and responsibility exist between headquarters and regional offices.

1.4.4 Airworthiness Inspection Division (AID) Responsibilities

All States should establish some form of airworthiness organization to meet the requirements set forth in the Convention and in Annexes 6 and 8. The organizational structure of an inspection organization within the CAA, hereinafter referred to as the AID, will vary depending upon the level and scope of aviation activity within the State and whether an AED has also been established.

The primary responsibilities of the AID should cover all matters concerning the continuing airworthiness of aircraft and should cover, at a minimum:

(a) continuing airworthiness of aircraft and parts thereof;

(b) approval of maintenance organizations;

(c) maintenance certification of air operators;

(d) where no separate licensing division exists, the approval of maintenance training organizations; and

(e) where no separate personnel licensing division exists, the licensing of aircraft maintenance personnel.

In States where an AED is not established, it may be necessary for the AID to be responsible for those engineering tasks associated with continuing airworthiness. These tasks may include evaluation and approval of repair and modification requests related to the continued operation of aircraft.

Part II Analysis, Study and Exercises

Words

accredit	[ə'kredɪt]	*v*. 把……归于,归因于;委派;信任,正式认可;授权
adverse	[əd'vɜːs]	*adj*. 不利的;相反的;敌对的
advisory	[əd'vaɪzərɪ]	*adj*. 顾问的;咨询的
		n. 警报
aerodrome	['erədrəʊm]	*n*. 小型飞机场
aerodynamic	[ˌerəʊdaɪ'næmɪk]	*adj*. (汽车等)流线型的,符合空气动力学原理的
aeronautical	[ˌerə'nɔːtɪkl]	*adj*. 航空(学)的;飞机设计制造的
allot	[ə'lɒt]	*v*. 分配,配给
amendment	[ə'mendmənt]	*n*. 修订;修正;美国宪法修正案
ample	['æmpl]	*adj*. 丰富的;足够的;宽敞的
annex	[ə'neks]	*v*. 附加(尤指对文件补充)
	['æneks]	*n*. 附属建筑;文件的附录; 附则; 附表; 附件
appropriate	[ə'prəʊprɪət]	*adj*. 适当的;恰当的;合适的
		v. 占用,拨出;适合于
assess	[ə'ses]	*v*. 评定;估价;对……征税;评估
attest	[ə'test]	*v*. 证明;证实;为……作证
audit	['ɔːdɪt]	*n*. 质量或标准的审查,检查
authorization	[ˌɔːθəraɪ'zeɪʃn]	*n*. 授权,认可;批准,委任
aviation	[ˌeɪvɪ'eɪʃn]	*n*. 航空;飞行术;飞机制造业
catalogue	['kætəlɒg]	*v*. 列入目录; 编入目录; 记载,登记(某人、某事等的详情)
cease	[siːs]	*v*. 停止;终止
collaborate	[kə'læbəreɪt]	*v*. 合作
commensurate	[kə'menʃərət]	*adj*. 相称的,相当的
commerce	['kɒmɜːs]	*n*. 贸易;商业;商务
confer	[kən'fɜː]	*v*. 商讨; 协商; 交换意见; 授予
convention	[kən'venʃn]	*n*. 大会;惯例;[计] 约定;[法] 协定;习俗; 常规
coordination	[kəʊˌɔːdɪ'neɪʃn]	*n*. 协作; 协调; 配合; 协调动作的能力
credentials	[krə'denʃlz]	*n*. 资格; 资历; 资格证书
defect	[dɪ'fekt]	*n*. 缺点; 缺陷
deferment	[dɪ'fɜːmənt]	*v*. 推迟; 延期; 延缓

deficiency	[dɪˈfɪʃnsɪ]	*n*. 缺乏；缺少；不足；缺点；缺陷
depiction	[dɪˈpɪkʃn]	*n*. 描绘；描画
designate	[ˈdezɪgneɪt]	*v*. 指定
designee	[ˌdezɪgˈniː]	*n*. 被指派者；被任命者
discharge	[dɪsˈtʃɑːdʒ]	*v*. 排出；卸货；准许(某人)离开；解雇 *n*. 排出(物)；放出(物)；流出(物)
discrepancy	[dɪsˈkrepənsɪ]	*n*. 差异；不符合
disposition	[ˌdɪspəˈzɪʃn]	*n*. 性情，性格；倾向，意愿；布置，排列；部署；分配权
disseminate	[dɪˈsemɪneɪt]	*v*. 散布，传播(信息、知识等)
division	[dɪˈvɪʒn]	*n*. 分开；分隔；分配；(分出来的)部分；部门；除(法)；分歧；不和；差异
dynamics	[daɪˈnæmɪks]	*n*. (人或事物)相互作用的方式，动态；力学；动力学；动力
effectiveness	[ɪˈfektɪvnəs]	*n*. 有效性
enactment	[ɪˈnæktmənt]	*n*. 制定，颁布；通过
enforcement	[ɪnˈfɔːsmənt]	*n*. 执行，实施；强制
enhance	[ɪnˈhɑːns]	*v*. 增强，提高
evacuation	[ɪˌvækjʊˈeɪʃ(ə)n]	*n*. 疏散；撤离
expertise	[ˌekspɜːˈtiːz]	*n*. 专门知识；专门技能；专长
framework	[ˈfreɪmwɜːk]	*n*. 框架，构架
grant	[grænt]	*v*. 准予，授予；同意，承认；给予，允许
hangar	[ˈhæŋə(r)]	*n*. 飞机库
helicopter	[ˈhelɪkɒptə]	*n*. 直升机
implement	[ˈɪmplɪment]	*v*. 实施
implementation	[ˌɪmpləmenˈteɪʃən]	*n*. 执行，履行；实施，贯彻；生效；完成；工具；仪器
incorporated	[ɪnˈkɔːpəreɪtɪd]	*adj*. 合并的 *v*. 合并；包含
integrity	[ɪnˈtegrətɪ]	*n*. 诚实正直；完整；完好
jurisdiction	[ˌdʒʊrɪsˈdɪkʃn]	*n*. 管辖权；管辖区域
justification	[ˌdʒʌstɪfɪˈkeɪʃn]	*n*. 说明
legislation	[ˌledʒɪsˈleɪʃn]	*n*. 法规；法律；立法；制定法律
liaison	[lɪˈeɪzn]	*n*. 连音；联络人；连读；联络
mandate	[ˈmændeɪt]	*n*. (政府或组织等经选举而获得的)授权；(政府的)任期；委托书；授权令 *v*. 强制执行；委托办理；授权
mandatory	[ˈmændətərɪ]	*adj*. 强制的；法定的；义务的 *n*. 受托者

means	[miːnz]	*n*. 方法
modification	[ˌmɒdɪfɪˈkeɪʃn]	*n*. 修改；改进；改变
navigation	[ˌnævɪˈgeɪʃn]	*n*. 导航；领航
nomination	[ˌnɒmɪˈneɪʃn]	*n*. 提名；推荐；任命；指派
obligation	[ˌɒblɪˈgeɪʃn]	*n*. 义务;职责;债务
operation	[ˌɒpəˈreɪʃn]	*n*. 手术；(有组织的)活动,行动;(包括许多部分的)企业,公司
overhaul	[ˌəʊvəˈhɔːl]	*n*. 检修 *v*. 彻底检修
oversight	[ˈəʊvəsaɪt]	*n*. 监督,照管;疏忽
pertinent	[ˈpɜːtɪnənt]	*adj*. 有关的；恰当的；相宜的
plating	[ˈpleɪtɪŋ]	*n*. 镀层(镀在金属上的其他金属薄层)；外层；(尤指)金属板护层 *v*. 电镀(尤指镀金、镀银)；为……加设护板；(用金属板等)覆盖
preamble	[ˈpriːæmbl]	*n*. 序言;绪论
promulgate	[ˈprɒmlgeɪt]	*v*. 公布;传播;发表
propeller	[prəˈpelə(r)]	*n*. [航][船] 螺旋桨;推进器
propulsion	[prəˈpʌlʃn]	*n*. 推动力；推进
prototype	[ˈprəʊtətaɪp]	*n*. 原型；雏形；最初形态
provision	[prəˈvɪʒn]	*n*. 供应,提供;(金钱)预备;供应量,提供物;条文,条款,规定
proximity	[prɒkˈsɪmətɪ]	*n*. (时间或空间)接近,邻近,靠近
pursuant	[pəˈsjuːənt]	*adj*. 依据的;追赶的;随后的 ad*v*. 根据;依照
ramp	[ræmp]	*n*. 斜坡;坡道
recommendation	[ˌrekəmenˈdeɪʃn]	*n*. 推荐;建议;推荐信
register	[ˈredʒɪstə]	*v*. 登记;注册
registry	[ˈredʒɪstrɪ]	*n*. 注册;登记处;挂号处;船舶的国籍
regulatory	[ˈregjələtərɪ]	*adj*. (对工商业)具有监管权的,监管的
relevant	[ˈreləvənt]	*adj*. 相关的;切题的;中肯的;有重大关系的;有意义的,目的明确的
remuneration	[rɪˌmjuːnəˈreɪʃn]	*n*. 酬金；薪水；报酬
renewal	[rɪˈnjuːəl]	*n*. 更新,恢复;复兴;补充;革新;续借;重申
revocation	[ˌrevəˈkeɪʃn]	*n*. 取消;撤回;废除
scrutiny	[ˈskruːtənɪ]	*n*. 仔细检查；认真彻底的审查
speciality	[ˌspeʃɪˈælətɪ]	*n*. 特产；特色菜；专业；专长

specification	[ˌspesɪfɪˈkeɪʃən]	n. 规格；规范；明细单；说明书
stipulate	[ˈstɪpjʊleɪt]	v. 规定；保证
		adj. 有托叶的
supervise	[ˈsuːpəvaɪz]	v. 监督；管理；指导；主管；照看
surveillance	[sɜːˈveɪləns]	n. 监督(对犯罪嫌疑人或可能发生犯罪的地方的)监视
suspension	[səˈspenʃn]	n. 悬浮；暂停；停职
tactful	[ˈtæktfl]	adj. 得体的；不得罪人的
territory	[ˈterətrɪ]	n. 领土
thereof	[ˌðeərˈɒv]	adv. 在其中；由此
undersigned	[ˌʌndəˈsaɪnd]	n. 签署人；签字人
undue	[ˌʌnˈduː]	adj. 不适当的，过分的
uniformity	[ˌjuːnɪˈfɔːmətɪ]	n. 相同性；不变性；统一性；一致性
validation	[ˌvælɪˈdeɪʃn]	n. 确认；批准；生效
viability	[ˌvaɪəˈbɪlətɪ]	n. 可行性
violation	[ˌvaɪəˈleɪʃ(ə)n]	n. 违反，违法
warrant	[ˈwɒrənt]	n. 执行令
		v. 使有必要
weld	[weld]	v. 焊接；熔接；锻接；使紧密结合；使连成整体
		n. 焊接点；焊接处

Phrases & Expressions

acceptable equivalent system	可接受的等效系统
accident and incident investigations	事故和事件调查
additional equipment	辅助设备；附加设备
aircraft maintenance	飞机维护；飞机维修
airworthiness advisory material	适航咨询材料
applied air and ground loads	施加的空气和地面载荷
appropriate corrective action	适当的纠正措施
as an alternative	作为一种替代品；作为一种替代
bear in mind	记住；考虑到
civil aviation regulation	民航条例
comparable to	比得上
competent independent organization	具备能力的独立机构
comply with	遵守，遵循
consistent with	与……一致
contracting state	缔约国

Convention on International Civil Aviation	《国际民用航空公约》
delegated task	委托任务
design approval	设计批准
discharge the responsibility	履行责任
engaged in	从事,参加
impending danger	迫在眉睫的危险
in accord with	符合,一致
in coordination with	与……协同,与……配合
in light of	根据;鉴于;从……观点
in regard to	关于;至于;在……方面
in the proximity of	在……附近
initial certificate of airworthiness	初始适航证
keep pace with	与……齐头并进(或并驾齐驱);与……步调一致
maintain consistency with	保持一致性
obliged to	被迫;被要求
on behalf of	代表
on-the-job training	在职培训
oversight system	监督系统
prior to	在……之前
quality assurance	质量保证
regardless of	不顾,不管
rendered valid	核准有效
safety oversight	安全监督
so as to	以便;以致
special flight permit	特殊飞行许可
specific technical speciality	特定技术专业
state of registry	登记国
surveillance and oversight function	监视和督查功能
technical competency	技术能力
to the greatest extent possible	尽最大可能
undersigned government	信守政府
with regard to	关于

Abbreviations

AD	Airworthiness Directive	适航指令
AED	Airworthiness Engineering Division	适航工程部门; 适航工程处; 适

		航评估部门
AID	Airworthiness Inspection Division	适航检查部门；适航检查科；适航审查部门
AMO	Approved Maintenance Organization	（主管部门）认可的维修单位
CAA	Civil Aviation Authority	民航局
DGCA	Director General of Civil Aviation	民航处处长
ICAO	International Civil Aviation Organization	国际民航组织
MCAI	Mandatory Continuous Airworthiness Information	强制性持续适航信息
MCM	Maintenance Control Manual	维护控制手册
MRB	Maintenance Review Board	维修审查委员会
MTOW	Maximum Take-Off Weight	最大起飞重量
RSOO	Regional Safety Oversight Organization	区域安全监督组织
SARPs	Standards and Recommended Practices	标准和建议方法
SB	Service Bulletin	服务通告
TCB	Type Certification Board	型号合格审定委员会

Sentence Comprehension

1. Article 54 of the Convention allows ICAO to issue Standards and Recommended Practices (SARPs) and to designate them as Annexes to the Convention.

   译文：公约第五十四条允许国际民航组织颁发标准和建议措施(SARPs)，并将其指定为公约的附件。

2. Although methods of discharging the foregoing State airworthiness responsibilities may vary, and in some cases may involve the transfer of certain tasks to authorized organizations or other States, such arrangements do not relieve the State of Registry from its overall responsibility.

   译文：虽然履行上述国家适航责任的方法可能有所不同，而且在某些情况下，可能涉及把某些任务转移到授权的组织或其他国家，但这样的安排并不解除登记国的总体责任。

3. Represent a well-balanced allocation of responsibility between the State and those persons or organizations conducting airworthiness-related activities (e. g. , provisions for surveillance and enforcement of the regulations).

   译文：在国家和那些执行与适航性有关活动的人或机构之间，职责分配须保持平衡(如监督条款和强制执行规定)。

4. The State of Registry should ensure that there is a system where information on faults, malfunctions, defects and other occurrences that may have adverse effects on the continuing airworthiness of the aircraft be transmitted to the organization

responsible for type design.

译文：登记国应确保有一个系统，将关于故障、失效、缺陷和其他可能对飞机继续适航产生不利影响的事件的资料转交给负责型号设计的组织。

5. Several Articles of the Convention make it clear that if standards adopted by a State are lower than those required by ICAO, aircraft, certificate holders, or persons with licenses or certificates may not be able to participate in international air navigation, except with the permission of the State or States whose territory is entered.

译文：公约中有几条清楚地表明，如果一个国家采用的标准低于国际民航组织所要求的标准，那么航空器、证书持有人或持有执照或证书的人可能不能参与国际空中航行，除非得到即将进入的国家的准许。

6. In those States where there are extensive aviation operations, manufacturing and maintenance, it will generally be necessary to establish within the CAA airworthiness organization an airworthiness engineering division (AED) and an airworthiness inspection division (AID).

译文：在有广泛的航空运行、制造和维修的国家，通常需要在民航当局适航性组织内部建立适航工程部门(AED)和适航检查部门(AID)。

7. The AED's activity will normally be directed toward design approvals, type certification, manufacturing approvals, evaluations of modifications and repairs proposed by manufacturers, aircraft owners, air operators, and AMOs, to correct deficiencies of aircraft already in service.

译文：适航工程部门的活动通常会针对设计审批、型号合格审定、制造审批，以及由制造商、航空器所有人、航空运营人和批准的维修机构(AMOs)为纠正现役航空器的缺陷而建议的改装和修理。

8. The organizational structure of an inspection organization within the CAA, hereinafter referred to as the airworthiness inspection division (AID), will vary depending upon the level and scope of aviation activity within the State and whether an AED has also been established.

译文：民航局内部的检查机构的组织结构——此后称适航检查部门(AID)，将根据国家内部的航空活动的不同水平、范围和是否已建立适航工程部门(AED)而有所不同。

9. Article 33 of the Convention requires that Contracting States recognize as valid the Certificate of Airworthiness issued or rendered valid by the State of Registry provided that the requirements under which the certificate was issued or rendered valid are equal to or above the minimum standards established by ICAO.

译文：公约第三十三条规定，签订公约的国家应承认飞行器注册国所签发或已生效的适航证书的有效性，但条件是签发证书或使证书有效的条件等于或超过国际民航组织规定的最低标准。

10. With the enactment of primary aviation legislation the State can develop and promulgate civil aviation regulations and practices, including airworthiness regulations consistent with the Annexes.

译文：随着主要航空立法的颁布，国家可以制定和颁布民用航空法规和惯例，包括与附件一致的适航规定。

11. The authority granting the exemptions should monitor such exemptions to ensure that continuation of the relief is warranted.

译文：准予豁免的当局应监督此类豁免，以确保豁免的持续性。

12. Regardless of the organizational arrangements established, the DGCA must bear in mind that the obligations of each State to comply with the provisions of Annexes 6 and 8 remain unchanged.

译文：无论已建立的组织如何安排，民航处处长必须牢记，每个国家都应遵守附件 6 和附件 8 规定的义务这一点均保持不变。

13. The CAAs should ensure they attract and retain technically competent personnel with the credibility and competence to interact with industry in an efficient and effective manner.

译文：民航局应该确保吸引和留住有技术能力的人员，这些技术人员具备以高效和有效的方式与行业互动的信誉和能力。

14. In States with significant manufacturing activity it may be useful to organize the AED along functional lines by organizing sections dealing with specific technical specialities, such as structures, propulsion, electrical and mechanical systems (including software), certification of production organizations, and surveillance and oversight functions.

译文：在具有重大制造活动的国家，通过具有专门技术专长（如结构、推进、电气和机械系统[包括软件]、生产组织认证，以及监督和监督职能）的部门，按照功能线组织 AED 可能是有用的。

15. The surveillance should cover compliance to approved or accepted procedures to obtain an accurate depiction of the day-to-day operation and also compliance with airworthiness requirements.

译文：监督视察应包括对于批准或接受的程序的符合，以获得对日常运行的准确描述，并符合适航要求。

16. When the AID has been notified in writing of corrective action, a follow-up visit should take place to verify correction of the discrepancies and compliance with the airworthiness requirements.

译文：当被以书面形式通知采取纠正措施时，适航检查部门应进行一次后续访问，以验证对不符点的纠正和对适航要求的符合性。

17. Means will need to be provided to ensure that the data continue to be available if the manufacturer ceases to support the internet documents, or the internet data are temporarily unavailable.

译文：如果制造商不再支持互联网文件，或者互联网数据暂时不可用，则需要提供确保文件仍然可用的方法。

Advanced Word Study

accredit [ə'kredɪt] *v.* 把……归于；归因于；委派；信任；正式认可；授权

Definitions:

① (formal) to believe that somebody is responsible for doing or saying something

② (specialist) to choose somebody for an official position

③ to officially approve something/somebody as being of an accepted quality or standard

Synonyms: credit, empower

Antonyms: doubt, suspect, dismiss

Usage examples:

① The discovery of distillation is usually *accredited* to the Arabs of the 11th century.

② The Arabs are usually *accredited* with the discovery of distillation.

③ The ICAO Regional Office *accredited* to the State may be of assistance to the DGCA in working out cooperative inspection arrangements.

catalogue ['kætəlɒg] *v./n.* 记载，登记（某人、某事等的详情）

Definitions:

① (*n.*) a complete list of items, for example of things that people can look at or buy

② (*v.*) to arrange a list of things in order in a catalogue; to record something in a catalogue

Synonyms: (*n.*) collection, directory, index, inventory, list

(*v.*) assemble, classify, compile, detail, enter

Usage examples:

① Interviews with the refugees *catalogue* a history of their lives.

② All records kept by the CAA should be *catalogued*, controlled and secured as required by State legislation, according to procedures defined by the CAA.

convention [kən'venʃn] *n.* 大会；[法] 惯例；[计] 约定；[法] 协定；习俗；常规

Definitions:

① the way in which something is done that most people in a society expect and consider to be polite or the right way to do it

② a large meeting of the members of a profession, a political party, etc.

Synonyms: custom; meeting, assembly

Antonyms: variation

Usage examples:

① Article 33 of the *Convention* requires that Contracting States recognize as valid the Certificate of Airworthiness issued or rendered valid by the State of Registry provided that the requirements under which the certificate was issued or rendered valid are equal to or above the minimum standards established by ICAO.

② Article 12 of the *Convention* points out that each Contracting State undertakes to keep its own regulations in these respects uniform, to the greatest extent possible, with those established from time to time under this Convention.

enforcement [ɪnˈfɔːsmənt] *n.* 执行,实施;强制

Definitions: the act of making people obey a particular law or rule

Synonyms: execution, implementation, administration, performance

Antonyms: volunteer

Usage examples:

① The doctors want stricter *enforcement* of existing laws.

② This process concerns the design and *enforcement* of policies.

grant [grænt] *v.* 准予,授予;同意,承认;给予,允许

Definitions:

① to agree to give somebody what they ask for, especially formal or legal permission to do something

② to admit that something is true, although you may not like or agree with it

Synonyms: permit

Antonyms: forbid, prohibit

Usage examples:

① The bank finally *granted* a $500 loan to me.

② The relief *granted* should be considered the exception and not the norm.

③ I was amazed that virtually all the things I took for *granted* up north just didn't happen in London.

jurisdiction [ˌdʒʊrɪsˈdɪkʃn] *n.* 管辖权;管辖区域

Definitions:

① the power, right, or authority to interpret and apply the law

② the authority of a sovereign power to govern or legislate

③ the limits or territory within which authority may be exercised

Synonyms: administration, authority, governance, government, rule

Usage examples:

① As a general policy, it is not desirable for CAA personnel to obtain technical qualifications from those entities under their direct regulatory *jurisdiction.*

② It must be clear that regardless of the arrangements made by a State, it is in no way relieved of the ultimate responsibility for the safe, regular and efficient conduct of aviation within its *jurisdiction.*

modification [ˌmɒdɪfɪˈkeɪʃn] *n*. 修改;改进;改变

Definitions:

① the act or process of changing something in order to improve it or make it more acceptable

② a change that is made

Synonyms: alteration, change, difference, review, revision

Antonyms: fixation, stabilization

Usage examples:

① Considerable *modification* of the existing system is needed.

② It might be necessary to make a few slight *modifications* to the design.

nomination [ˌnɒmɪˈneɪʃn] *n*. 提名;推荐;任命;指派

Definitions:

① the act, process, or an instance of nominating

② the state of being nominated

Synonyms: proposal, suggestion, recommendation, submission, choice

Usage examples:

① We expect him to get the *nomination.*

② Membership is by *nomination* only.

③ The novel earned a *nomination* for the National Book Award.

④ The film received five Academy Award *nominations.*

navigation [ˌnævɪˈgeɪʃn] *n*. 导航;领航

Definitions:

① the skill or the process of planning a route for a ship or other vehicle and taking it there

② the movement of ships or aircraft

Synonyms: pilotage, piloting

Antonyms: missing one's way, losing one's way, getting lost

Usage examples:

① They will be helped by the very latest in *navigation* aids.

② Overweight aircraft carrying extra fuel or *navigation* equipment.

obliged　　[əˈblaɪdʒd]　　*v.* 被迫;被要求　　*adj.* 感激的;感谢的

Definitions:

① used when you are expressing thanks or asking politely for something, to show that you are grateful to somebody

② forced by law

Synonyms: force, necessitate, request

Antonyms: volunteer

Usage examples:

① I'm much *obliged* to you for helping us.

② The storm got worse and worse. Finally, I was *obliged* to abandon the car and continue on foot.

③ We called up three economists to ask how to eliminate the deficit and they *obliged* with very straightforward answers.

④ ICAO is *obliged* to immediately notify all other States of the difference which exists between one or more aspects of the Standard and the corresponding national practice of the notifying State.

operation　　[ˌɒpəˈreɪʃn]　　*n.* 手术;(有组织的)活动,行动;(包括许多部分的)企业,公司

Definitions:

① an organized activity that involves several people doing different things

② a business or company involving many parts

Synonyms: action, company, business

Usage examples:

① The technical library should also be able to reference all documents issued by ICAO relating to the *operation* and airworthiness of aircraft.

② If the CAA is not able to obtain the necessary data when the manufacturer ceases its activity, it should assess if the continued safe *operation* of affected aircraft is still achievable.

overhaul　　[ˌəʊvəˈhɔːl]　　*n.* 检修　*v.* 彻底检修

Definitions: an examination of a machine or system, including doing repairs on it or making changes to it

Synonyms: renovation, redevelopment, modernization

Antonyms: wreck, ruin

Usage examples: They had ensured the airplane was under *overhaul* a year ago.

pertinent [ˈpɜːtɪnənt] *adj*. 有关的;恰当的;相宜的

Definitions: appropriate to a particular situation

Synonyms: relevant, related, appropriate

Antonyms: irrelevant

Usage examples:

① Please keep your comments *pertinent* to the topic under discussion.

② The AED should evaluate proposals *pertinent* to manufacturing aspects of the design, repair and modification of an aircraft or its parts to ensure conformity with CAA specifications.

prototype [ˈprəʊtətaɪp] *n*. 原型;雏形;最初形态

Definitions: the first design of something from which other forms are copied or developed

Synonyms: case, example, illustration, instance, representative, sample, specimen

Usage examples:

① The car is presently at the *prototype* stage.

② The team is developing a *prototype* for a digital compact camera.

③ The AED should issue special flight permits for aircraft that do not meet applicable airworthiness requirements, but are capable of safe flight (e. g. *prototype* aircraft or production flight tests).

④ Inspect *prototype* aircraft, test specimens and test installation, as necessary.

register [ˈredʒɪstə] *v*. 登记;注册 *n*. 登记表;注册簿

Definitions:

① to record your/somebody's/something's name on an official list

② to give opinion publicly

Synonyms: enroll, inscribe, list

Antonyms: delist

Usage examples:

① Article 31 of the Convention provides that the Certificate of Airworthiness is issued or rendered valid by the State in which the aircraft is *registered.*

② In the development of national airworthiness regulations and rules, the State of Registry is responsible to make certain that every aircraft on its *register* conforms to the approved type design in accordance with the airworthiness code it has adopted or accepted for that class of aircraft.

surveillance [sɜːˈveɪləns] *n*. 监督;(对犯罪嫌疑人或可能发生犯罪的地方的)监视;视察

Definitions: close watch kept over someone or something

Synonyms: care, charge, guidance, oversight, regulation, supervision

Usage examples:

① Government *surveillance* of international affairs.

② The bank robbery was recorded by *surveillance* video cameras.

③ The State's technical personnel performing certification and *surveillance* functions on behalf of the State should be at least as qualified as the personnel to be inspected or supervised.

④ In addition to the importance of technical competency in performing certification, inspection and *surveillance* functions, it is critical that inspectors possess a high degree of integrity, be impartial in carrying out their tasks, be tactful, have a good understanding of human nature and possess good communication skills.

transmit [trænzˈmɪt] *v.* 传送;输送;发射;播送;传播;传染;传(热、声等);透(光等);使通过

Definitions:

① to send an electronic signal, radio or television broadcast, etc.

② (formal) to pass something from one person to another

③ (specialist) to allow heat, light, sound, etc. to pass through

Synonyms: communicate, conduct, convey, give, spread, transfer

Antonyms: accept, receive, hold, keep, retain, catch

Usage examples: Along with relevant airworthiness responsibilities of the State of Design, State of Manufacture, State of Registry, and all Contracting States, the chapter also includes requirements for States to *transmit* information.

violate [ˈvaɪəleɪt] *v.* 违反,违法

Definitions:

① (formal) to go against or refuse to obey a law, an agreement, etc.

② (formal) to disturb or not respect somebody's peace, privacy, etc.

③ to damage or destroy a holy or special place

Synonyms: breach, break, fracture, offend

Antonyms: comply (with), conform (to), follow, meet, mind, obey, observe

Usage examples: Article 12 of the Convention not only requires uniformity of such regulations with the Convention but obligates States to ensure the prosecution of all persons *violating* relevant regulations.

Exercises for Self-Study

Ⅰ. Translate Sentences

1. The aircraft should satisfy the airworthiness requirements of Annex 8, but may

not be usable for a specific operational task without meeting the additional requirements of Annex 6.

2. Primary aviation legislation should contain provisions to establish a civil aviation administration that proactively supervises and regulates civil aviation activities.
3. Adopting another State's regulations has some advantages but should be considered only after ensuring the other State's regulations are consistent with the ICAO Standards.
4. The new inspector should be teamed with an experienced inspector who will ensure the on-the-job training is performed and documented.
5. Article 31 of the Convention provides that the Certificate of Airworthiness is issued or rendered valid by the State in which the aircraft is registered.
6. With well documented regulations in place a State may grant an exception or exemption to a regulation provided there exists a mechanism within the States criteria to establish that exception or exemption.
7. If an exemption is granted it should be accompanied by conditions and limitations, including a time limitation to the exemption.
8. It is also recognized that a State or group of States may elect to discharge their responsibilities through agreements with a regional safety oversight organization or agency.
9. It is essential that the agreements clearly define the respective functions each party is to perform, so as to ensure that all obligations of the States are fully discharged.
10. This assists the airworthiness personnel in determining whether or not mandatory modifications, inspections and repairs approved by the State of Design are appropriately carried out.

Ⅱ. Multiple Choice Questions

1. Article 54 of the Convention allows ICAO to issue Standards and Recommended Practices (SARPs) and to ________ them as Annexes to the Convention.

   A. discharge B. designate C. collaborate D. enhance

2. Annex 8 provides that the State of Design transmits information which it has found necessary for the continuing airworthiness of aircraft and provides notification of the ________ or ________ of a type certificate.

   A. legislation B. suspension C. revocation D. convention

3. It is essential that the agreements clearly define the respective functions each party is to perform, so as to ensure that all obligations of the States are fully________.

   A. charged B. changed C. discharged D. completed

4. In addition to the importance of technical competency in performing certification,

inspection and ________ functions, it is critical that inspectors possess a high degree of integrity, be impartial in carrying out their tasks, be tactful, have a good understanding of human nature and possess good communication skills.

A. watch B. monitor C. glare D. surveillance

5. When the physical size of the State is large and the level of aviation activity is relatively high, it may be necessary to establish regional offices in the ________ of the aviation industry.

A. near B. proximity C. approach D. close

6. Furthermore, specialized areas such as non-destructive ________, ________, ________, software quality assurance and special manufacturing processes should be considered.

A. inspection B. inspecting C. planting

D. plating E. warding F. welding

7. The surveillance should cover compliance to approved or accepted procedures to obtain an accurate________ of the day-to-day operation and also compliance with airworthiness requirements.

A. depiction B. violation C. navigation D. deficiency

8. All records kept by the CAA should be________, controlled and secured as required by State legislation.

A. conferred B. catalogued C. division D. operated

9. Article 38 of the Convention provides when the State finds it impracticable to comply with any international standard or procedure, or to bring its regulations or practices ________ any international standard or procedure after amendment, that it notifies ICAO of differences between its own practice and that established by the ICAO Standards.

A. satisfy B. comply with C. in accord with D. follow

10. Be able to make national airworthiness compliance determinations ________ manufacturing operations and be able to ensure that the organization conforms to their quality control programme.

A. regard as B. with regard to C. refer to D. depend on

11. The AED's activity will normally be directed toward design approvals, type certification, manufacturing approvals, evaluations of modifications and repairs proposed by manufacturers, ________, ________, and AMOs, to correct deficiencies of aircraft already in service.

A. aircraft owners B. air operators

C. aircraft supplier D. aircraft manufacturer

12. The AID should ________ national and international levels on matters relating to the regulations and technical matters concerning airworthiness.

A. confer at B. confer to C. work at D. control

Ⅲ. Fill in the Blanks

1. Annex 8 outlines a ________ of airworthiness standards for the design and manufacture of large aeroplanes, helicopters, small aeroplanes, engines and propellers.
2. The State is urged to take immediate and appropriate actions ________ the State oversight system.
3. Primary aviation ________ is essential to an effective safety oversight programme.
4. Primary aviation legislation should contain provisions to establish a civil aviation administration that proactively ________ and regulates civil aviation activities.
5. In those States which do not have an aviation manufacturing industry, the airworthiness organization within the CAA will be mainly concerned with inspection, ________ and approval functions.
6. Granting or ________ of aircraft noise certification and issuance of attestation of noise certification.
7. It is preferable that personnel also possess aeronautical licenses, certificates and/or aeronautical degrees ________ with their job responsibilities.
8. The ICAO Regional Office ________ to the State may be of assistance in working out cooperative inspection arrangements among States in the region.
9. The AED's activity will normally be directed ________ design approvals, type certification, manufacturing approvals, evaluations of modifications and repairs proposed by manufacturers, aircraft owners, air operators, and AMOs, to correct deficiencies of aircraft already in service.
10. Flights after a modification or repair or during a process of applying for a supplemental type certificate, evacuation of aircraft from areas of ________.
11. Investigate possible violation of the national air law or regulation ________ airworthiness and take appropriate enforcement action.

Ⅳ. Grammar/Logical Mistakes Correction

1. Article 31 of the Convention provides that the Certificate of Airworthiness is issued or rendered valid by the State which the aircraft is registered.
2. Article 33 of the Convention requires that Contracting States recognize as valid the Certificate of Airworthiness issued or rendered valid by the State of Registry provided that the requirements are less than the minimum standards established by ICAO.

3. The promulgation of civil airworthiness regulations and procedures is necessary for the CAA to effectively discharge its official airworthiness responsibilities and carry out those duties associated with the design, manufacture, operation, maintenance of aircraft and aeronautical products and the certification of maintenance personnel.
4. A State may grant an exception or exemption to a regulation provided there exists a mechanism in the States criteria to establish that exception or exemption.
5. If an exemption is granted it should be accompanied by conditions and limitations, including a place limitation to the exemption.
6. The agreements clearly define the respective functions each party is to perform, as to ensure that all obligations of the States are fully discharged.
7. The State's technical personnel performing certification and surveillance functions behalf of the State should be at least as qualified as the personnel to be inspected or supervised.
8. The AED should develop and evaluate changes in engineering standards, procedures and practices to reflect on current requirements and limitations and keeping pace with changes in aviation technology.
9. The CAA will also need to keep files from each aircraft registered in the State.

Ⅴ. True or False

1. Primary aviation legislation is essential to an effective safety oversight programme. The extent of the CAA's authority and empowerment and that of its Director General of Civil Aviation (DGCA) needn't be reflected in the legislation thereby giving the CAA a solid foundation in a legal document.
2. The State of Registry, upon receipt from the State of Design, either adopts the information or will assess the information and take appropriate action.
3. The airworthiness organization should identify and resolve regulatory problems associated with continuing airworthiness and establish appropriate general and technical regulations, policies and procedures.
4. An AID inspector generally should have the relevant knowledge, background and appropriate experience related to aircraft continuing airworthiness management.
5. The AED should issue special flight permits for aircraft that meet applicable airworthiness requirements, but are not capable of safe flight (e. g. prototype aircraft or production flight tests).
6. The surveillance programme only includes unannounced surveillance visits of the certificate holders.
7. The AID should ensure all surveillance that is performed is properly documented

and referenced and retained for future audits.

8. For CAA, all Internet correspondence requiring a response should be given a due date for a reply.

Ⅵ. Questions to Answer

1. What are the annexes related to the airworthiness of aircraft?
2. Who has the responsibility for the issuance of approvals?
3. Which two departments are generally included in CAA?
4. How is an AID inspector selected?
5. When can the technical on-the-job training of airworthiness inspectors finish?
6. When any unsafe situation occurs, who should disseminate the corresponding information?
7. In States where an AED is not established, what shall be responsible for those engineering tasks associated with continuing airworthiness?

Ⅶ. Questions for Discussion

1. What is the purpose of enacting primary aviation legislation by the State?
2. Do airworthiness organizations play a role in balancing airworthiness?
3. Why is compliance with airworthiness regulations not optional?
4. The CAA needs to register each aircraft in the State, what shall be included in the registration file?
5. What is surveillance programme of the aviation industry performed by the State?
6. What is the purpose of the on-the-job training of airworthiness inspectors?
7. Do you agree that ICAO can replace CAA and regulate civil aviation on its own?

Chapter 2 Airworthiness Standards: Performance

Part Ⅰ Regulatory Document Text Example

[from Federal Aviation Regulations FAR 25 25.101—25.125]

2.1 General

(a) Unless otherwise prescribed, airplanes must meet the applicable performance requirements of this subpart for ambient atmospheric conditions and still air.

(b) The performance, as affected by engine power or thrust, must be based on the following relative humidities: (1) for turbine engine powered airplanes, a relative humidity of: (ⅰ) 80 percent, at and below standard temperatures; and (ⅱ) 34 percent, at and above standard temperatures plus 50 ℉. Between these two temperatures, the relative humidity must vary linearly; (2) for reciprocating engine powered airplanes, a relative humidity of 80 percent in a standard atmosphere.

(c) The performance must correspond to the propulsive thrust available under the particular ambient atmospheric conditions, the particular flight condition, and the relative humidity.

The available propulsive thrust must correspond to engine power or thrust, not exceeding the approved power or thrust less: (1) installation losses; and (2) the power or equivalent thrust absorbed by the accessories and services appropriate to the particular ambient atmospheric conditions and the particular flight condition.

(d) Unless otherwise prescribed, the applicant must select the takeoff, en route, approach, and landing configurations for the airplane.

(e) The airplane configurations may vary with weight, altitude, and temperature, to the extent they are compatible with the operating procedures.

(f) Unless otherwise prescribed, in determining the accelerate-stop distances, takeoff flight paths, takeoff distances, and landing distances, changes in the airplane's configuration, speed, power, and thrust, must be made in accordance with procedures established by the applicant for operation in service.

(g) Procedures for the execution of balked landings and missed approaches associated with the conditions prescribed in § 25.119 and 25.121 must be established.

(h) The procedures must: (1) be able to be consistently executed in service by crews of average skill; (2) use methods or devices that are safe and reliable; and (3) include allowance for any time delays, in the execution of the procedures, that may reasonably be expected in service.

(i) The accelerate-stop and landing distances prescribed in § 25.109 and 25.125, respectively, must be determined with all the airplane wheel brake assemblies at the fully worn limit of their allowable wear range.

2.2 Stall Speed

(a) The reference stall speed, V_{SR}, is a calibrated airspeed defined by the applicant. V_{SR} may not be less than a 1-g stall speed.

(b) V_{CLMAX} is determined with: (1) engines idling, or, if that resultant thrust causes an appreciable decrease in stall speed, not more than zero thrust at the stall speed; (2) propeller pitch controls (if applicable) in the takeoff position; (3) the airplane in other respects (such as flaps and landing gear) in the condition existing in the test or performance standard in which V_{SR} is being used; (4) the weight used when V_{SR} is being used as a factor to determine compliance with a required performance standard; (5) the center of gravity position that results in the highest value of reference stall speed; and (6) the airplane trimmed for straight flight at a speed selected by the applicant, but not less than $1.13V_{SR}$ and not greater than $1.3V_{SR}$.

(c) Starting from the stabilized trim condition, apply the longitudinal control to decelerate the airplane so that the speed reduction does not exceed one knot per second.

(d) When a device that abruptly pushes the nose down at a selected angle of attack (e.g., a stick pusher) is installed, the reference stall speed, V_{SR}, may not be less than 2 knots or 2 percent, whichever is greater, above the speed at which the device operates.

2.3 Takeoff

(a) The takeoff speeds, the accelerate-stop distance, the takeoff path, and the takeoff distance and takeoff run, must be determined: (1) at each weight, altitude, and ambient temperature within the operational limits selected by the applicant; and (2) in the selected configuration for takeoff.

(b) No takeoff made to determine the data required by this section may require exceptional piloting skill or alertness.

(c) The takeoff data must be based on: (1) in the case of land planes and amphibians: (ⅰ) smooth, dry and wet, hard-surfaced runways; and (ⅱ) at the option of the applicant, grooved or porous friction course wet, hard-surfaced runways; (2) smooth water, in the case of seaplanes and amphibians; and (3) smooth, dry snow, in the case of skiplanes.

(d) The takeoff data must include, within the established operational limits of the airplane, the following operational correction factors: (1) not more than 50 percent of nominal wind components along the takeoff path opposite to the direction of takeoff, and not less than 150 percent of nominal wind components along the takeoff path in the direction of takeoff; (2) effective runway gradients.

2.4 Takeoff Speeds

(a) V_1 must be established in relation to V_{EF} as follows: (1) V_{EF} is the calibrated airspeed at which the critical engine is assumed to fail. (2) V_{EF} must be selected by the applicant, but may not be less than V_{MCG}.

V_1, in terms of calibrated airspeed, is selected by the applicant; however, V_1 may not be less than V_{EF} plus the speed gained with critical engine inoperative during the time interval between the instant at which the critical engine is failed, and the instant at which the pilot recognizes and reacts to the engine failure, as indicated by the pilot's initiation of the first action (e. g., applying brakes, reducing thrust, deploying speed brakes) to stop the airplane during accelerate-stop tests.

(b) V_{2MIN}, in terms of calibrated airspeed, may not be less than: (1) 1.13 V_{SR} for: (ⅰ) two-engine and three-engine turbopropeller and reciprocating engine powered airplanes; and (ⅱ) turbojet powered airplanes without provisions for obtaining a significant reduction in the one-engine-inoperative power-on stall speed; (2) 1.08 V_{SR} for: (ⅰ) turbopropeller and reciprocating engine powered airplanes with more than three engines; and (ⅱ) turbojet powered airplanes with provisions for obtaining a significant reduction in the one-engine-inoperative power-on stall speed; and (3) 1.10 times V_{MC}.

(c) V_2, in terms of calibrated airspeed, must be selected by the applicant to provide at least the gradient of climb but may not be less than: (1) V_{2MIN}; (2) V_R plus the speed increment attained before reaching a height of 35 feet above the takeoff surface; and (3) a speed that provides the maneuvering capability.

(d) V_{MU} is the calibrated airspeed at and above which the airplane can safely lift off the ground, and continue the takeoff. V_{MU} speeds must be selected by the applicant throughout the range of thrust-to-weight ratios to be certificated. These speeds may be established from free air data if these data are verified by ground takeoff tests.

(e) V_R, in terms of calibrated airspeed, must be selected in accordance with the conditions:

(1) V_R may not be less than: (i) V_1; (ii) 105 percent of V_{MC}; (iii) the speed that allows reaching V_2 before reaching a height of 35 feet above the takeoff surface; or (iv) a speed that, if the airplane is rotated at its maximum practicable rate, will result in a V_{LOF} of not less than 110 percent of V_{MU} in the all-engines operating condition and not less than 105 percent of V_{MU} determined at the thrust-to-weight ratio corresponding to the one-engine -inoperative condition.

(2) For any given set of conditions (such as weight, configuration, and temperature), a single value of V_R, obtained in accordance with this paragraph, must be used to show compliance with both the one-engine-inoperative and the all-engines-operating takeoff provisions.

(3) It must be shown that the one-engine inoperative takeoff distance, using a rotation speed of 5 knots less than V_R, does not exceed the corresponding one engine inoperative takeoff distance using the established V_R.

(4) Reasonably expected variations in service from the established takeoff procedures for the operation of the airplane (such as over-rotation of the airplane and out-of-trim conditions) may not result in unsafe flight characteristics or in marked increases in the scheduled takeoff distances.

(f) V_{LOF} is the calibrated airspeed at which the airplane first becomes airborne.

(g) V_{FTO}, in terms of calibrated airspeed, must be selected by the applicant to provide at least the gradient of climb, but may not be less than: (1) $1.18V_{SR}$; and (2) a speed that provides the maneuvering capability.

2.5 Takeoff Path

(a) The takeoff path extends from a standing start to a point in the takeoff at which the airplane is 1,500 feet above the takeoff surface, or at which the transition from the takeoff to the en route configuration is completed and V_{FTO} is reached, whichever point is higher. In addition: (1) the takeoff path must be based on the procedures; (2) the airplane must be accelerated on the ground to V_{EF}, at which point the critical engine must be made inoperative and remain inoperative for the rest of the takeoff; and (3) after reaching V_{EF}, the airplane must be accelerated to V_2.

(b) During the acceleration to speed V_2, the nose gear may be raised off the ground at a speed not less than V_R. However, landing gear retraction may not be begun until the airplane is airborne.

(c) During the takeoff path determination in accordance with paragraphs (a) and (b) of this section: (1) the slope of the airborne part of the takeoff path must be positive at each point; (2) the airplane must reach V_2 before it is 35 feet above the takeoff surface and must continue at a speed as close as practical to, but not less than V_2, until it is 400 feet above the takeoff surface; (3) at each point along the takeoff path, starting at the point at which the airplane reaches 400 feet above the takeoff surface, the available gradient of climb may not be less than: (ⅰ) 1.2 percent for two-engine airplanes; (ⅱ) 1.5 percent for three-engine airplanes; and (ⅲ) 1.7 percent for four-engine airplanes; and (4) except for gear retraction and propeller feathering, the airplane configuration may not be changed, and no change in power or thrust that requires action by the pilot may be made, until the airplane is 400 feet above the takeoff surface.

(d) The takeoff path must be determined by a continuous demonstrated takeoff or by synthesis from segments.

If the takeoff path is determined by the segmental method: (1) the segments must be clearly defined and must be related to the distinct changes in the configuration, power or thrust, and speed; (2) the weight of the airplane, the configuration, and the power or thrust must be constant throughout each segment and must correspond to the most critical condition prevailing in the segment; (3) the flight path must be based on the airplane's performance without ground effect; and (4) the takeoff path data must be checked by continuous demonstrated takeoffs up to the point at which the airplane is out of ground effect and its speed is stabilized, to ensure that the path is conservative relative to the continuous path.

The airplane is considered to be out of the ground effect when it reaches a height equal to its wing span.

2.6 Takeoff Distance and Takeoff Run

(a) Takeoff distance on a dry runway is the greater of: (1) the horizontal distance along the takeoff path from the start of the takeoff to the point at which the airplane is 35 feet above the takeoff surface, determined for a dry runway; or (2) 115 percent of the horizontal distance along the takeoff path, with all engines operating, from the start of the takeoff to the point at which the airplane is 35 feet above the takeoff surface, as determined by a procedure.

(b) Takeoff distance on a wet runway is the greater of: (1) the takeoff distance on a dry runway determined in accordance with paragraph (a) of this section; or (2) the horizontal distance along the takeoff path from the start of the takeoff to the point at which the airplane is 15 feet above the takeoff surface, achieved in a manner consistent with the achievement of V_2 before reaching 35 feet above the takeoff surface, determined for a wet runway.

(c) If the takeoff distance does not include a clearway, the takeoff run is equal to the takeoff distance. If the takeoff distance includes a clearway, the takeoff run on a dry runway is the greater of: (i) the horizontal distance along the takeoff path from the start of the takeoff to a point equidistant between the point at which V_{LOF} is reached and the point at which the airplane is 35 feet above the takeoff surface, as determined for a dry runway; or (ii) 115 percent of the horizontal distance along the takeoff path, with all engines operating, from the start of the takeoff to a point equidistant between the point at which V_{LOF} is reached and the point at which the airplane is 35 feet above the takeoff surface, determined by a procedure.

2.7 Climb: General

Compliance with the requirements must be shown at each weight, altitude, and ambient temperature within the operational limits established for the airplane and with the most unfavorable center of gravity for each configuration.

2.8 Landing Climb: All-Engines-Operating

In the landing configuration, the steady gradient of climb may not be less than 3.2 percent, with: (1) the engines at the power or thrust that is available eight seconds after initiation of movement of the power or thrust controls from the minimum flight idle to the go-around power or thrust setting; and (2) a climb speed of not more than V_{REF}.

2.9 Climb: One-Engine-Inoperative

(a) Takeoff: landing gear extended. In the critical takeoff configuration existing along the flight path (between the points at which the airplane reaches V_{LOF} and at which the landing gear is fully retracted) and in the configuration but without ground effect, the steady gradient of climb must be positive for two-engine airplanes, and not less than 0.3 percent for three-engine airplanes or 0.5 percent for four-engine airplanes, at V_{LOF} and with: (1) the critical engine inoperative and the remaining engines at the power or thrust available when retraction of the landing gear is begun unless there is a more critical power operating condition existing later along the flight path but before the point at which the landing gear is fully retracted; and (2) the weight equal to the weight existing when retraction of the landing gear is begun.

(b) Takeoff: landing gear retracted. In the takeoff configuration existing at the point of the flight path at which the landing gear is fully retracted, and in the configuration but without ground effect, the steady gradient of climb may not be less than 2.4 percent for two-engine airplanes, 2.7 percent for three-engine airplanes, and 3.0 percent for four-engine airplanes, at V_2 and with: (1) the critical engine inoperative, the remaining engines at the takeoff power or thrust available at the time the landing gear is fully retracted unless there is a more critical power operating condition existing later along the flight path but before the point where the airplane reaches a height of 400 feet above the takeoff surface; and (2) the weight equal to the weight existing when the airplane's landing gear is fully retracted.

(c) Final takeoff. In the en route configuration at the end of the takeoff path, the steady gradient of climb may not be less than 1.2 percent for two-engine airplanes, 1.5 percent for three-engine airplanes and 1.7 percent for four-engine airplanes, at V_{FTO} and with: (1) the critical engine inoperative and the remaining engines at the available maximum continuous power or thrust; and (2) the weight equal to the weight existing at the end of the takeoff path.

(d) Approach. In a configuration corresponding to the normal all-engines-operating procedure in which V_{SR} for this configuration does not exceed 110 percent of the V_{SR} for the related all-engines-operating landing configuration, the steady gradient of climb may not be less than 2.1 percent for two-engine airplanes, 2.4 percent for three-engine airplanes, and 2.7 percent for four engine airplanes, with: (1) the critical engine inoperative, the remaining engines at the go-around power or thrust setting; (2) the maximum landing weight; (3) a climb speed established in connection with normal landing procedures, but not more than 1.4 V_{SR}; and (4) landing gear retracted.

2.10 En route Flight Paths

(a) For the en route configuration, the flight paths of this section must be determined at each weight, altitude, and ambient temperature, within the operating limits established for the airplane. The variation of weight along the flight path, accounting for the progressive consumption of fuel and oil by the operating engines, may be included in the computation. The flight paths must be determined at any selected speed, with: (1) the most unfavorable center of gravity; (2) the critical engines inoperative; (3) the remaining engines at the available maximum continuous power or thrust; and (4) the means for controlling the engine cooling air supply in the position that provides adequate cooling in the hot-day condition.

(b) The one-engine-inoperative net flight path data must represent the actual climb performance diminished by a gradient of climb of 1.1 percent for two-engine airplanes, 1.4 percent for three-engine airplanes, and 1.6 percent for four-engine airplanes.

(c) For three- or four-engine airplanes, the two-engine-inoperative net flight path data must represent the actual climb performance diminished by a gradient of climb of 0.3 percent for three-engine airplanes and 0.5 percent for four-engine airplanes.

2.11 Landing

(a) The horizontal distance necessary to land and to come to a complete stop (or to a speed of approximately 3 knots for water landings) from a point 50 feet above the landing surface must be determined (for standard temperatures, at each weight, altitude, and wind within the operational limits established by the applicant for the airplane) as follows:

(1) The airplane must be in the landing configuration.

(2) A stabilized approach, with a calibrated airspeed of V_{REF}, must be maintained down to the 50 feet height. V_{REF} may not be less than (i) 1.23 V_{SR0}; (ii) V_{MCL}; and (iii) a speed that provides the maneuvering capability.

(3) Changes in configuration, power or thrust, and speed, must be made in accordance with the established procedures for service operation.

(4) The landing must be made without excessive vertical acceleration, tendency to bounce, nose over, ground loop, porpoise, or water loop.

(5) The landings may not require exceptional piloting skill or alertness.

(b) For landplanes and amphibians, the landing distance on land must be determined on a level, smooth, dry, hard-surfaced runway. In addition: (1) the pressures on the wheel braking systems may not exceed those specified by the brake manufacturer; (2) the brakes may not be used so as to cause excessive wear of brakes or tires; and (3) means other than wheel brakes may be used if that means: (i) is safe and reliable; (ii) is used so that consistent results can be expected in service; and (iii) is such that exceptional skill is not required to control the airplane.

(c) For seaplanes and amphibians, the landing distance on water must be determined on smooth water.

(d) For skiplanes, the landing distance on snow must be determined on smooth, dry, snow.

(e) The landing distance data must include correction factors for not more than 50 percent of the nominal wind components along the landing path opposite to the direction of landing, and not less than 150 percent of the nominal wind components along the landing path in the direction of landing.

(f) If any device is used that depends on the operation of any engine, and if the landing distance would be noticeably increased when a landing is made with that engine inoperative, the landing distance must be determined with that engine inoperative unless the use of compensating means will result in a landing distance not more than that with each engine operating.

Part II Analysis, Study and Exercises

Words

accessory [əkˈsesərɪ] *n*. 附件；配件；附属物；(衣服的)配饰；从犯；同谋；帮凶
adj. 辅助的；副的

airborne [ˈeəbɔːn] *adj*. 空气传播的；空降的

airspeed [ˈeəspiːd] *n*. 空速

alertness [əˈlɜːtnəs] *n*. 警觉,清醒,机警

ambient [ˈæmbɪənt] *adj*. 周围环境的；周围的；产生轻松氛围的

appendix [əˈpendɪks] *n*. 阑尾；(书、文件的)附录

applicable [əˈplɪkəbl] *adj*. 适用的；合适的

assembly [əˈsemblɪ] *n*. 立法机构；会议；议会；集会；(统称)集会者；(全校师生的)晨会,朝会

atmospheric [ˌætməsˈferɪk] *adj*. 大气的;大气层的;令人激动的;使人动感情的

calibrated [ˈkælɪbreɪtɪd] *adj*. 校正的

clearway [ˈklɪəweɪ] *n*. 净空道

climb [klaɪm] *v*. 爬升

compatible [kəmˈpætəbl] *adj*. 可共用的；兼容的；可共存的；(因志趣等相投而)关系好的,和睦相处的

compliance [kəmˈplaɪəns] *n*. 服从；顺从；遵从

configuration [kənˌfɪgəˈreɪʃn] *n*. 布局；结构；构造；格局；形状；(计算机的)配置；形态

crew [kruː] *n*. (轮船、飞机等上面的)全体工作人员；技术人

员团队；专业团队

v. 当(尤指船上的)工作人员

demonstrate [ˈdemənstreɪt] *v*. 证明；证实；论证；说明；表达；表露；表现；显露；示范；演示

equivalent [ɪˈkwɪvələnt] *adj*. (价值、数量、意义、重要性等)相等的，相同的

n. 相等的东西；等量；对应词

exceptional [ɪkˈsepʃənl] *adj*. 杰出的；优秀的；卓越的；异常的；特别的；罕见的

execution [ˌeksɪˈkjuːʃn] *n*. 处决；实行；执行；实施；表演；(乐曲的)演奏；(艺术品的)制作

extent [ɪkˈstent] *n*. 程度；限度；大小；面积；范围

factor [ˈfæktə] *n*. 因素；要素；因子；因数

gradient [ˈgreɪdɪənt] *n*. (尤指公路或铁路的)坡度，斜率，倾斜度；(温度、压力等的)变化率，梯度变化曲线

grooved [gruːvd] *adj*. 有沟的；有槽的

v. 在……上开出沟(或槽等)；跟着流行乐跳舞；过得快活

horizontal [ˌhɔːrɪˈzɒntl] *adj*. 水平的

humidity [hjuːˈmɪdətɪ] *n*. (空气中的)湿度；湿热；高温潮湿

idle [ˈaɪdl] *v*. 慢车；挂空挡，空转

adj. 懈怠的；懒惰的；闲置的；漫无目的的；无效的；无用的

increment [ˈɪŋkrəmənt] *n*. 增量；增加

initiation [ɪˌnɪʃɪˈeɪʃn] *n*. 开始；创始；发起；引发

level [ˈlevl] *adj*. 平的；平坦的

linearly [ˈlɪnɪəlɪ] *adv*. (成)直线地；线性地

nominal [ˈnɒmɪnl] *adj*. 名义上的；有名无实的；很小的；象征性的；名词性的；名词的

n. 名词性词

performance [pəˈfɔːməns] *n*. 表现；性能；业绩；工作情况；做；执行；履行

porous [ˈpɔːrəs] *adj*. 多孔的；透水的；透气的

prescribe [prɪˈskraɪb] *v*. 给……开(药)；让……采用(疗法)；开(处方)；规定；命令；指示

prevailing [prɪˈveɪlɪŋ] *adj*. 普遍的；盛行的；流行的；(指风)一地区常刮的

		v. 普遍存在；盛行；流行；被接受；战胜；压倒；(尤指长时间斗争后)战胜，挫败
propeller	[prəˈpelə(r)]	*n*. 螺旋桨(飞机或轮船的推进器)
reciprocating	[rɪˈsɪprəkeɪtɪŋ]	*v*. 回报；回应；沿直线往复移动
retraction	[rɪˈtrækʃn]	*n*. 撤销；收回；拉回
segment	[ˈsegmənt]	*n*. 部分；份；片；段；(柑橘、柠檬等的)瓣；弓形；圆缺 *v*. 分割；划分；分段
setting	[ˈsetɪŋ]	*n*. 设置
standby	[ˈstændbaɪ]	*n*. 后备人员；备用物品 *adj*. (机票、音乐会门票等) 最后时刻出售的
steady	[ˈstedɪ]	*adj*. (发展、增长等)稳步的，持续的，匀速的；稳定的；恒定的；平稳的；稳固的 *v*. 使平稳；稳住；恢复平稳；使平静；使冷静；使镇定
subpart	[sʌbˈpɑːt]	*n*. 次元件；子部分；次组件；子节；分部分
synthesis	[ˈsɪnθəsɪs]	*n*. 综合；结合；综合体；(物质在动植物体内的)合成；(人工的)合成
temperature	[ˈtemprətʃə]	*n*. 温度；气温；体温 *adj*. 大气的；大气层的；令人激动的；使人动感情的
tendency	[ˈtendənsɪ]	*n*. 倾向；偏好；性情；趋势；趋向；(政党内的)极端派别
thrust	[θrʌst]	*v*. 猛推；冲；搡；挤；塞；刺；戳 *n*. (论据、政策等的)要点，要旨，重点；猛推；刺；戳；插；(发动机推动飞机、火箭等的)推力，驱动力
transition	[trænˈzɪʃn]	*n*. 过渡；转变；变革；变迁 *v*. 经历转变过程；过渡
unfavorable	[ʌnˈfeɪvərəbl]	*adj*. 不利的；相反的；令人不快的；不吉利的
variation	[ˌveərɪˈeɪʃn]	*n*. (数量、水平等的)变化，变更，变异；变异的东西；变种；变体
vertical	[ˈvɜːtɪkl]	*adj*. 竖的；垂直的；直立的；纵向的 *n*. 垂直线；垂直位置

Phrases & Expressions

accelerate-stop distance	中断起飞距离

account for	对……负有责任;对……做出解释;说明……的原因;导致;比例占
all-engines-operating procedure	全发工作操作程序
ambient atmospheric conditions	环境大气条件
appropriate to	将(某物)分配给……
balked landing	中断着陆
be the greater of	两者中较大
compensating means	补偿手段
correspond to	相发于;对应于
critical engine	关键发动机;临界发动机
demonstrated takeoff	试飞演示
determined horizontal distance	确定的水平距离
dry/wet runway	干/湿跑道
en route configuration	航路布局;航路形态;航线配置
established operational limits	既定操作限制范围
exceptional piloting skill	杰出的驾驶技能;特殊的驾驶技巧
gradient of climb	爬升梯度
ground effect	地面效应
horizontal distance	水平距离
in accordance with	依照;与……一致;符合;和……一致
landing distance	着陆距离
landing gear retraction	起落架收缩
lateral control	横向操纵
longitudinal control	纵向操纵
maneuvering capability	机动能力;操纵性
missed approach	复飞
net flight path data	净飞行轨迹数据
nose gear	前起落架
one-engine-inoperative	单发停车
operational correction factors	操作修正因素
operational limits	操作极限;运行限制
performance requirements	性能要求
piloting skill	驾驶技术
progressive consumption	累进消耗;逐渐消耗
propeller feathering	螺旋桨顺桨
relative humidity	相对湿度
remaining engine	剩余发动机

stabilized approach	稳定进场；稳定进近
stall speed	失速速度
still air	静止空气
takeoff distance	起飞距离
takeoff path	起飞航迹
takeoff run	起飞滑跑；起飞滑跑距离
takeoff surface	起飞面
to the extent	在某种程度上
turbine engine	涡轮发动机；涡轮机
two-engine airplane	双引擎飞机
vary with	随……而变化
wing span	翼展，翼层

Abbreviations

V_{FTO}	Final Takeoff Speed	最终起飞速度
V_{EF}	Engine Failure Speed	发动机失效速度
V_2	Takeoff Safety Speed	飞机的安全起飞速度
V_R	Rotation Speed	飞机的抬轮速度/抬头速度

Sentence Comprehension

1. Unless otherwise prescribed, airplanes must meet the applicable performance requirements of this subpart for ambient atmospheric conditions and still air.

   译文：除非另有规定，飞机必须满足本子部分适用于环境大气条件和静止空气的性能要求。

2. The airplane in other respects (such as flaps and landing gear) in the condition existing in the test or performance standard in which VSR is being used.

   译文：该飞机在其他方面(例如襟翼、起落架和冰积聚)处于使用 VSR 的试验或性能标准所具有的状态。

3. No takeoff made to determine the data required by this section may require exceptional piloting skill or alertness.

   译文：为确定本节要求的数据而进行的起飞，可能不需要特殊的驾驶技能或机敏度。

4. Reasonably expected variations in service from the established takeoff procedures for the operation of the airplane (such as over-rotation of the airplane and out-of-trim conditions) may not result in unsafe flight characteristics or in marked increases in the scheduled takeoff distances established in accordance with § 25.113(a).

译文：飞机运行起飞程序之外的合理的预期服务变化（如飞机的过旋和失稳情况），可能不会导致不安全飞行特性，或不会导致第 25.113(a)条规定的预定起飞距离的显著增加。

5. At each point along the takeoff path, starting at the point at which the airplane reaches 400 feet above the takeoff surface, the available gradient of climb may not be less than 1.2 percent for two-engine airplanes.

译文：在起飞航迹上的每一点，从飞机到达起飞面以上 400 英尺的那一点开始，双引擎飞机可用的爬升梯度可能不小于 1.2%。

6. The engines at the power or thrust that is available eight seconds after initiation of movement of the power or thrust controls from the minimum flight idle to the go-around power or thrust setting.

译文：动力或推力启动后 8 秒内可用的发动机功率（推力）控制从最小飞行怠速到复飞动力或推力设置。

7. The critical engine inoperative and the remaining engines at the power or thrust available when retraction of the landing gear is begun in accordance with § 25.111 unless there is a more critical power operating condition existing later along the flight path but before the point at which the landing gear is fully retracted.

译文：依据第 25.111 条，起落架完全收起时，临界发动机停车，其他发动机功率或推力可用，除非随后沿飞行航迹在起落架完全收起之前，存在更临界的动力装置运转状态。

8. The weight equal to the weight existing when the airplane's landing gear is fully retracted, determined under § 25.111.

译文：重量与第 25.111 条确定的起落架完全收起时的重量相等。

9. The variation of weight along the flight path, accounting for the progressive consumption of fuel and oil by the operating engines, may be included in the computation.

译文：在计算中也应考虑由发动机燃油的累进消耗导致的飞机的重量随航道改变而发生的变化。

10. If any device is used that depends on the operation of any engine, and if the landing distance would be noticeably increased when a landing is made with that engine inoperative, the landing distance must be determined with that engine inoperative unless the use of compensating means will result in a landing distance not more than that with each engine operating.

译文：如果使用依赖发动机操作的设备，且如果着陆时发动机不起作用，着陆距离会明显增加，那么着陆距离必须在引擎不起作用时确定，除非使用补偿手段可以使着陆距离不高于引擎运转时的距离。

11. Except for gear retraction and propeller feathering, the airplane configuration

may not be changed, and no change in power or thrust that requires action by the pilot may be made, until the airplane is 400 feet above the takeoff surface.

译文：除了起落架收缩和螺旋桨顺桨外，飞机的构型不能改变，并且在飞机距起飞面 400 英尺高之前，不得改变需要飞行员操作的动力或推力。

12. Takeoff distance on a dry runway is the greater of: (1) the horizontal distance along the takeoff path from the start of the takeoff to the point at which the airplane is 35 feet above the takeoff surface, determined under § 25.111 for a dry runway; or (2) 115 percent of the horizontal distance along the takeoff path, with all engines operating, from the start of the takeoff to the point at which the airplane is 35 feet above the takeoff surface, as determined by a procedure consistent with § 25.111.

    译文：干跑道上的起飞距离是以下两者中较大的值：(1)根据第 25.111 条的规定，从起飞开始到飞机离起飞面 35 英尺时的水平距离；(2)根据第 25.111 条的规定，在所有发动机工作的情况下，从起飞时到飞机离起飞面 35 英尺处的水平距离的 115%。

13. The prescribed reduction in climb gradient may be applied as an equivalent reduction in acceleration along that part of the takeoff flight path at which the airplane is accelerated in level flight.

    译文：规定的爬升梯度的减小可以作为飞机在水平飞行中加速时沿起飞飞行路径的那一部分加速度的等效减小。

Advanced Word Study

approach [əˈprəʊtʃ]

v. (在距离或时间上)靠近，接近；接洽；建议；要求；(在数额、水平或质量上)接近

n. (待人接物或思考问题的)方式，方法，态度；(距离和时间上的)靠近，接近；接洽；建议；要求

Definitions:

① to come near to somebody/something in distance or time

② to start dealing with a problem, task, etc. in a particular way

③ the part of an aircraft's flight immediately before landing

Synonyms: method, way, system, mechanism, medium

Antonyms: retreat

Usage examples:

① We heard the sound of an *approaching* car.

② She *approached* the bank for a loan.

③ What's the best way of *approaching* this problem?

④ We have been *approached* by a number of companies that are interested in our product.

⑤ An airplane begins the final *approach* to the runway.

⑥ A stabilized *approach*, with a calibrated airspeed of VREF, must be maintained down to the 50 feet height.

correspond [ˌkɒrəˈspɒnd] *v*. 相当于；对应于

Definitions:

① to be the same as or match something

② to be similar to or the same as something else

③ to be in conformity or agreement

④ to be equivalent or parallel

Synonyms: accord, consist, fall in with

Antonyms: differ (from), disagree (with)

Usage examples:

① Your account of events does not *correspond* with hers.

② The British job of Lecturer *corresponds* roughly to the U. S. Associate Professor.

③ In some countries, the role of president *corresponds* to that of prime minister.

④ The performance must *correspond* to the propulsive thrust available under the particular ambient atmospheric conditions, the particular flight condition, and the relative humidity specified in paragraph (b) of this section.

⑤ The weight of the airplane, the configuration, and the power or thrust must be constant throughout each segment and must *correspond* to the most critical condition prevailing in the segment.

demonstrate [ˈdemənstreɪt] *v*. 证明;证实;论证;说明;表达;表露;表现;显露;示范;演示

Definitions:

① to show something clearly by giving proof or evidence

② to show by your actions that you have a particular quality, feeling or opinion

③ to show and explain how something works or how to do something

④ to take part in a public meeting or march, usually as a protest or to show support for something

⑤ to make known (something abstract) through outward signs

Synonyms: establish, prove, show, display, substantiate

Antonyms: disprove

Usage examples:

① You need to *demonstrate* more self-control.

② Her job involves *demonstrating* new educational software.

③ We want to *demonstrate* our commitment to human rights.

④ Let me *demonstrate* to you how it works.

⑤ The babysitter's actions during the emergency *demonstrate* beyond doubt her general dependability.

⑥ The takeoff path must be determined by a continuous *demonstrated* takeoff or by synthesis from segments.

⑦ The test loads spectrum applied to the structure should be *demonstrated* to be conservative when compared to the usage expected in service.

⑧ The applicant should *demonstrate* that his quality system (e. g. design, process control, and material standards) ensures the scatter in fatigue properties is controlled, and that the design of the fatigue critical areas of the part account for the material scatter.

determine [dɪˈtɜːmɪn] *v.* 查明;测定;准确算出;决定;形成;支配;影响;确定;裁决;安排

Definitions:

① to discover the facts about something; to calculate something exactly

② to make something happen in a particular way or be of a particular type

③ to officially decide and/or arrange something

④ to decide definitely to do something

Synonyms: arbitrate, decide, judge, rule (on), settle, resolve, discover, ascertain

Usage examples:

① An inquiry was set up to *determine* the cause of the accident.

② We set out to *determine* exactly what happened that night.

③ It was *determined* that she had died of natural causes.

④ Age and experience will be *determining* factors in our choice of candidate.

⑤ A date for the meeting has yet to be *determined.*

⑥ The court *determined* that the defendant should pay the legal costs.

⑦ The demand for a product *determines* its price.

⑧ The takeoff flight path shall be considered to begin 35 feet above the takeoff surface at the end of the takeoff distance *determined* in accordance with § 25.113(a) or (b), as appropriate for the runway surface condition.

⑨ The flight paths must be *determined* at any selected speed.

execution　　[ˌeksɪˈkjuːʃn]　　*n.* 处决；实行；执行；实施；表演；（乐曲的）演奏；（艺术品的）制作

Definitions:

① the act of killing somebody, especially as a legal punishment

② (formal) the act of doing a piece of work, performing a duty, or putting a plan into action

③ (formal) skill in performing or making something, such as a piece of music or work of art

④ (law) the act of following the instructions in a legal document, especially those in somebody's will

Synonyms: performance, fulfillment

Usage examples:

① He faced *execution* by hanging for murder.

② He had failed in the *execution* of his duty.

③ Her *execution* of the piano piece was perfect.

④ The lawers are proceeding with the *execution* of her mother's will.

⑤ Procedures for the *execution* of balked landings and missed approaches associated with the conditions prescribed in § 25. 119 and 25. 121(d) must be established.

⑥ Include allowance for any time delays, in the *execution* of the procedures, that may reasonably be expected in service.

gradient　　[ˈgreɪdɪənt]　　*n.* 倾向；意愿；趋向；趋势；倾斜度

Definitions:

① the rate of regular or graded/ascent or descent: inclination

② a part sloping upward or downward

③ change in the value of a quantity (such as temperature, pressure, or concentration) with change in a given variable and especially per unit distance in a specified direction

Synonyms: inclination, incline, lean, pitch, slope

Usage examples:

① The field was on a slight *gradient*.

② The road has a fairly steep *gradient*.

③ The net takeoff flight path data must be determined so that they represent the actual takeoff flight paths reduced at each point by a *gradient* of climb.

④ The one-engine-inoperative net flight path data must represent the actual climb performance diminished by a *gradient* of climb of 1. 1 percent for two-engine airplanes.

manner [ˈmænə] *n.* 方式;方法;举止;态度;礼貌;礼仪

Definitions:

① (formal) the way that something is done or happens

② the way that somebody behaves towards other people

③ behaviour that is considered to be polite in a particular society or culture

④ the habits and customs of a particular group of people

Synonyms: etiquette

Usage examples:

① The *manner* in which the decision was announced was extremely regrettable.

② His *manner* was polite but cool.

③ It is bad *manners* to talk with your mouth full.

④ The second is the promise that when legal disputes do arise, they will be resolved in a timely *manner* — before, not after, the targeted party has been bled of precious time and resources.

⑤ For your last enhancement, you will introduce another function in the same *manner* as above.

prescibe [prɪˈskraɪb] *v.* 给……开(药);让……采用(疗法);开(处方);规定;命令;指示

Definitions:

① (of a doctor) to tell somebody to take a particular medicine or have a particular treatment; to write a prescription for a particular medicine, etc.

② (of a person or an organization with authority) to say what should be done or how something should be done

Synonyms: define, lay down, specify

Usage examples:

① The doctor may be able to *prescribe* you something for that cough.

② The *prescribed* form must be completed and returned to this office.

③ Police regulations *prescribe* that an officer's number must be clearly visible.

④ Unless otherwise *prescribed*, airplanes must meet the applicable performance requirements of this subpart for ambient atmospheric conditions and still air.

⑤ Procedures for the execution of balked landings and missed approaches associated with the conditions *prescribed* in § 25. 119 and 25. 121(d) must be established.

procedure [prəˈsiːdʒə(r)] *n.* (正常)程序,手续,步骤;(商业、法律或政治上的)程序;手术

Definitions:

① a way of doing something, especially the usual or correct way

② the official or formal order or way of doing something, especially in business, law or politics

Synonyms: course, operation, proceeding, process

Usage examples:

① The *procedure* for logging on to the network usually involves a password.

② The agency admitted that agreeing to this was not at all standard operating *procedure* and is against our policies, and the assignment on those terms should have been declined.

③ The organization has established emergency/safety/disciplinary *procedures.*

④ He is familiar with export *procedure.*

⑤ This subpart prescribes the *procedure* for sloving the issue of airworthiness certificates.

⑥ 115 percent of the horizontal distance along the takeoff path, with all engines operating, from the start of the takeoff to the point at which the airplane is 35 feet above the takeoff surface, as determined by a *procedure* consistent with § 25.111.

Exercises for Self-Study

Ⅰ. Translate Sentences

1. The airplane configurations may vary with weight, altitude, and temperature, to the extent they are compatible with the operating procedures required by paragraph (f) of this section.
2. The takeoff data must include, within the established operational limits of the airplane, the following operational correction factors.
3. The weight of the airplane, the configuration, and the power or thrust must be constant throughout each segment and must correspond to the most critical condition prevailing in the segment.
4. The flight path must be based on the airplane's performance without ground effect.
5. For the en-route configuration, the flight paths prescribed in paragraphs (b) and (c) of this section must be determined at each weight, altitude, and ambient temperature, within the operating limits established for the airplane.
6. For landplanes and amphibians, the landing distance on land must be determined on a level, smooth, dry, hard-surfaced runway.
7. The consumption of resources is divided into several grades. Different grades are charged according to different charging standards.

8. The critical engine of a multi-engine, fixed-wing aircraft is the one whose failure would result in the most adverse effects on the aircraft's handling and performance.
9. The airplane is considered to be out of the ground effect when it reaches a height equal to its wing span.
10. The agency admitted that agreeing to this was not at all standard operating procedure and is against our policies, and the assignment on those terms should have been declined.

Ⅱ. Multiple Choice Questions

1. The ________ data must be checked by continuous ________ up to the point at which the airplane is out of ground effect and its speed is stabilized.
   A. takeoff path　B. VR　C. landing gear retraction
   D. demonstrated takeoff　E. ground effect
2. If the takeoff distance does not include a ________, the ________ is equal to the takeoff distance.
   A. takeoff run　B. takeoff path　C. takeoff surface
   D. horizontal distance　E. clearway
3. For the ________ the flight paths must be determined at each weight, altitude, and ambient temperature, within the operating limits established for the airplane.
   A. airplane　B. en route configuration
   C. airport runway　D. avionics instruments
4. The variation of weight along the flight path, accounting for the ________ of fuel and oil by the ________, may be included in the computation.
   A. amount　B. operating engines
   C. progressive consumption　D. specifications
5. The weight of the airplane, the ________, and the power or thrust must be constant throughout each segment and must correspond to the most critical condition prevailing in the segment.
   A. speed　B. angle of attack
   C. configuration　D. vertical speed/climb rate

Ⅲ. Fill in the Blanks

1. Unless otherwise ________, airplanes must meet the applicable performance requirements for ambient atmospheric conditions and still air.
2. The airplane configurations may vary with weight, altitude, and temperature, to the extent they are ________ with the operating procedures.
3. The weight of the airplane, the configuration, and the power or thrust must be

constant throughout each segment and must correspond to the most critical condition ________ in the segment.

4. The landing distance must be determined with that engine inoperative unless the use of ________ means will result in a landing distance not more than that with each engine operating.
5. The landings may not require ________ piloting skill or alertness.
6. The power or ________ thrust absorbed by the accessories and services appropriate to the particular ambient atmospheric conditions and the particular flight condition.
7. No takeoff made to determine the data may require exceptional piloting skill or ________.
8. The performance, as affected by engine power or thrust, must be based on the following relative ________.
9. Unless otherwise prescribed, the applicant must select the takeoff, en route, approach, and landing ________ for the airplane.
10. Be able to be consistently executed in service by ________ of average skill.
11. The takeoff path must be determined by a continuous demonstrated takeoff or by synthesis from ________.

Ⅳ. Grammar/Logical Mistakes Correction

1. The horizontal distance along the takeoff path from the start of the takeoff to the point at which the airplane is 15 feet above the takeoff surface, achieved in a manner consistent in the achievement of V_2 before reaching 35 feet above the takeoff surface.
2. Compliance with the requirements must be shown at each weight, altitude, and ambient temperature with the operational limits, with the most unfavorable center of gravity for each configuration.
3. The landings may don't require exceptional piloting skill or alertness.
4. For seaplanes and amphibians, the landing distance above water must be determined.
5. The critical engine of a multi-engine, fixed-wing aircraft is the one that failure would result in the most adverse effects on the aircraft's handling and performance.

Ⅴ. True or False

1. Takeoff made to determine the data required by this section may require exceptional piloting skill or alertness.
2. The takeoff path extends from a standing start to a point in the takeoff at which the airplane is 1,500 feet above the takeoff surface, or at which the transition

from the takeoff to the en route configuration is completed.

3. If the takeoff distance includes a clearway, the takeoff run is equal to the takeoff distance.
4. For landplanes and amphibians, the landing distance on land must be determined on a level, smooth, dry, hard-surfaced runway.
5. The flight path must be based on the airplane's performance including ground effect.

Ⅵ. Questions to Answer

1. How does the relative humidity vary?
2. May landings need exceptional piloting skill or alertness according to the requirements?
3. What are the flight stages for an aircraft to complete a typical flight?
4. When to consider an airplane going out of the ground effect?
5. What is the relationship between Takeoff Distance and Takeoff Run?
6. What is the definition of "Takeoff Flight Path"?

Ⅶ. Questions for Discussion

1. What requirements should aircraft performance meet?
2. When someone needs to use any compensating means, does it mean one's initial work went wrong?
3. What are the conditions that the flight paths are determined at?
4. Other than wheel brakes, what kind of means could be used for landplanes and amphibians?
5. Why do you think the regulated takeoff distance contains the distance used to climb up to 35 feet altitude?

Chapter 3 Airworthiness Standards: Structure

Part I Regulatory Document Text Example

[from Acceptable Means of Compliance AMC No. 2 to CS 25.301(b), AMC 25.307, AMC 25.491, AMC 25.571(a), (b) and (e)]

3.1 Flight Load Validation

(a) Measurements:

Flight load measurements (e.g., through application of strain gauges, pressure belts, accelerometers) may include: pressures / air loads /net shear, bending and torque on primary aerodynamic surfaces; flight mechanics parameters necessary to correlate the analytical model with flight test results; high lift devices loads and positions; primary control surface hinge moments and positions; unsymmetric loads on the empennage (due to roll/yaw maneuvers and buffeting); local strains or response measurements in cases where load calculations or measurements are indeterminate or unreliable.

(b) Conditions:

In the conduct of flight load measurements, conditions used to obtain flight loads may include: pitch maneuvers including wind-up turns, pull-ups and push-downs (e.g., for wing and horizontal stabiliser maneuvering loads); stall entry or buffet onset boundary conditions (e.g., for horizontal stabiliser buffet loads); yaw maneuvers including rudder inputs and steady sideslips; roll maneuvers.

Some flight load conditions are difficult to validate by flight load measurements, simply because the required input (e.g., gust velocity) cannot be accurately controlled or generated. Therefore, these type of conditions need not be flight tested. Also, in general, failures, malfunctions or adverse conditions are not subject to flight tests for the purpose of flight loads validation.

3.2 Proof of Structure

3.2.1 Definitions

(a) Detail. A structural element of a more complex structural member (e.g., joints, splices, stringers, stringer run-outs, or access holes).

(b) Sub Component. A major three-dimensional structure which can provide complete structural representation of a section of the full structure (e. g. , stub-box, section of a spar, wing panel, wing rib, body panel, or frames).

(c) Component. A major section of the airframe structure (e. g. , wing, body, fin, horizontal stabiliser) which can be tested as a complete unit to qualify the structure.

(d) Full Scale. Dimensions of test article are the same as design; fully representative test specimen (not necessarily complete airframe).

(e) New Structure. Structure for which behaviour is not adequately predicted by analysis supported by previous test evidence. Structure that utilises significantly different structural design concepts such as details, geometry, structural arrangements, and load paths or materials from previously tested designs.

(f) Similar New Structure. Structure that utilises similar or comparable structural design concepts such as details, geometry, structural arrangements, and load paths concepts and materials to an existing tested design.

(g) Derivative/Similar Structure. Structure that uses structural design concepts such as details, geometry, structural arrangements, and load paths, stress levels and materials that are nearly identical to those on which the analytical methods have been validated.

(h) Previous Test Evidence. Testing of the original structure that is sufficient to verify structural behaviour.

3.2.2 Introduction

The structure must be shown to comply with the strength and deformation requirements of Subpart C of CS -25. This means that the structure must:

(ⅰ) be able to support limit loads without detrimental permanent deformation, and

(ⅱ) be able to support ultimate loads without failure.

This implies the need of a comprehensive assessment of the external loads, the resulting internal strains and stresses, and the structural allowables.

CS 25.307 requires compliance for each critical loading condition. Compliance can be shown by analysis supported by previous test evidence, analysis supported by new test evidence or by test only. As compliance by test only is impractical in most cases, a large portion of the substantiating data will be based on analysis.

There are a number of standard engineering methods and formulas which are known to produce acceptable, often conservative results especially for structures where load paths are well defined. Those standard methods and formulas, applied with a good understanding of their limitations, are considered reliable analyses when showing compliance. Conservative assumptions may be considered in assessing whether or not an analysis may be accepted without test substantiation.

The application of methods such as Finite Element Method or engineering formulas to complex structures in modern aircraft is considered reliable only when validated by full scale tests (ground and/or flight tests). Experience relevant to the product in the utilisation of such methods should be considered.

3.2.3 Classification of Structure

(a) The structure of the product should be classified into one of the following three categories:

- New Structure;
- Similar New Structure;
- Derivative/Similar Structure.

(b) Justifications should be provided for classifications other than New Structure. Elements that should be considered are: (ⅰ) the accuracy/conservatism of the analytical methods, and (ⅱ) comparison of the structure under investigation with previously tested structure.

Considerations should include, but are not limited to the following: external loads (bending moment, shear, torque, etc.); internal loads (strains, stresses, etc.); structural design concepts such as details, geometry, structural arrangements, load paths; materials; test experience (load levels achieved, lessons learned); deflections; deformations; extent of extrapolation from test stress levels.

3.2.4 Certification Approaches

The following certification approaches may be selected:

(a) Analysis, supported by new strength testing of the structure to limit and ultimate load. This is typically the case for New Structure.

Substantiation of the strength and deformation requirements up to limit and ultimate loads normally requires testing of sub-components, full scale components or full scale tests of assembled components (such as a nearly complete airframe). The entire test program should be considered in detail to assure the requirements for strength and deformation can be met up to limit load levels as well as ultimate load levels.

Sufficient limit load test conditions should be performed to verify that the structure meets the deformation requirements and to provide validation of internal load distribution and analysis predictions for all critical loading conditions.

Because ultimate load tests often result in significant permanent deformation, choices will have to be made with respect to the load conditions applied. This is usually based on the number of test specimens available, the analytical static strength margins of safety of the structure and the range of supporting detail or sub-component tests. An envelope approach may be taken, where a combination of different load cases is applied, each one of which is critical for a different section of the structure.

These limit and ultimate load tests may be supported by detail and sub-component tests that verify the design allowables (tension, shear, compression) of the structure and often provide some degree of validation for ultimate strength.

(b) Analysis validated by previous test evidence and supported with additional limited testing. This is typically the case for Similar New Structure.

The extent of additional limited testing (number of specimens, load levels, etc.) will depend upon the degree of change.

For example, if the changes to an existing design and analysis necessitate extensive changes to an existing test-validated finite element model (e. g. different rib spacing) additional testing may be needed. Previous test evidence can be relied upon whenever practical.

These additional limited tests may be further supported by detail and sub-component tests that verify the design allowables (tension, shear, compression) of the structure and often provide some degree of validation for ultimate strength.

(c) Analysis, supported by previous test evidence. This is typically the case for Derivative/ Similar Structure.

Justification should be provided for this approach by demonstrating how the previous static test evidence validates the analysis and supports showing compliance for the structure under investigation.

For example, if the changes to the existing design and test-validated analysis are evaluated to assure they are relatively minor and the effects of the changes are well understood, the original tests may provide sufficient validation of the analysis and further testing may not be necessary. For example, if a weight increase results in higher loads along with a corresponding increase in some of the element thickness and fastener sizes, and materials and geometry (overall configuration, spacing of structural members, etc.) remain generally the same, the revised analysis could be considered reliable based on the previous validation.

(d) Test only.

Sometimes no reliable analytical method exists, and testing must be used to show compliance with the strength and deformation requirements. In other cases it may be elected to show compliance solely by tests even if there are acceptable analytical methods. In either case, testing by itself can be used to show compliance with the strength and deformation requirements of CS-25 Subpart C. In such cases, the test load conditions should be selected to assure all critical design loads are encompassed.

If tests only are used to show compliance with the strength and deformation requirements for single load path structure which carries flight loads (including pressurisation loads), the test loads must be increased to account for variability in material properties. In lieu of a rational analysis, for metallic materials, a factor of 1.15 applied to the limit and ultimate flight loads may be used. If the structure has multiple load paths, no material correction factor is required.

3.2.5 Interpretation of Data

The interpretation of the substantiation analysis and test data requires an extensive review of: the representativeness of the loading; the instrumentation data; comparisons with analytical methods; representativeness of the test article(s); test set-up (fixture, load introductions); load levels and conditions tested; test results.

Testing is used to validate analytical methods except when showing compliance by test only. If the test results do not correlate with the analysis, the reasons should be identified and appropriate action taken. This should be accomplished whether or not a test article fails below ultimate load.

Should a failure occur below ultimate load, an investigation should be conducted for the product to reveal the cause of this failure. This investigation should include a review of the test specimen and loads, analytical loads, and the structural analysis. This may lead to adjustment in analysis/modelling techniques and/or part redesign and may result in the need for additional testing. The need for additional testing to ensure ultimate load capability, depends on the degree to which the failure is understood and the analysis can be validated by the test.

3.3 Taxi, Take-Off and Landing Roll

3.3.1 Purpose

This AMC sets forth acceptable methods of compliance with the provisions of CS-25 dealing with the certification requirements for taxi, take-off and landing roll design loads. Guidance information is provided for showing compliance, relating to structural design for aeroplane operation on paved runways and taxi-ways normally used in commercial operations. Other methods of compliance with the requirements may be acceptable.

3.3.2 Background

All paved runways and taxi-ways have an inherent degree of surface unevenness, or roughness. This is the result of the normal tolerances of engineering standards required for construction, as well as the result of events such as uneven settlement and frost heave. In addition, repair of surfaces on an active runway or taxi-way can result in temporary ramped surfaces. Many countries have developed criteria for runway surface roughness. The ICAO standards are published in ICAO Annex 14.

Several approaches had been taken by different manufacturers in complying with the noted regulations. If dynamic effects due to rigid body modes or airframe flexibility during taxi were not considered critical, some manufacturers used a simplified static analysis where a static inertia force was applied to the aeroplane using a load factor of 2. 0 for single axle gears or 1. 7 for multiple axle gears. The lower 1. 7 factor was justified based on an assumption that there was a load alleviating effect resulting from rotation of the beam, on which the forward and aft axles are attached, about the central pivot point on the strut. The static load factor approach was believed to encompass any dynamic effects and it had the benefit of a relatively simple analysis.

3. 4 Damage Tolerance and Fatigue Evaluation of Structure

3. 4. 1 Introduction

(a) The contents of this AMC are considered by the Agency in determining compliance with the damage-tolerance and fatigue requirements of CS 25. 571.

Damage-tolerance design is required, unless it entails such complications that an effective damage-tolerant structure cannot be achieved within the limitations of geometry, inspectability, or good design practice. Under these circumstances, a design that complies with the fatigue evaluation (safe-life) requirements is used. Typical examples of structure that might not be conducive to damage-tolerance design are landing gear, engine mounts, and their attachments.

(b) Typical Loading Spectra Expected in Service. The loading spectrum should be based on measured statistical data of the type derived from government and industry load history studies and, where insufficient data are available, on a conservative estimate of the anticipated use of the aeroplane. The principal loads that should be considered in establishing a loading spectrum are flight loads (gust and maneuver), ground loads (taxiing, landing impact, turning, engine runup, braking, and towing) and pressurisation loads. The development of the loading spectrum includes the definition of the expected flight plan which involves climb, cruise, descent, flight times, operational speeds and altitudes, and the approximate time to be spent in each of the operating regimes. Operations for crew training, and other pertinent factors, such as the dynamic stress characteristics of any flexible structure excited by turbulence, should also be considered. For pressurised cabins, the loading spectrum should include the repeated application of the normal operating differential pressure, and the super-imposed effects of flight loads and external aerodynamic pressures.

(c) Analyses and Tests. Unless it is determined from the foregoing examination that the normal operating stresses in specific regions of the structure are of such a low order that serious damage growth is extremely improbable, repeated load analyses or tests should be conducted on structure representative of components or sub-components of the wing, control surfaces, empennage, fuselage, landing gear, and their related primary attachments. Test specimens should include structure representative of attachment fittings, major joints, changes in section, cutouts, and discontinuities. Any method used in the analyses should be supported, as necessary, by test or service experience.

3.4.2 Damage-Tolerance (Fail-Safe) Evaluation

(a) General. The damage-tolerance evaluation of structure is intended to ensure that should serious fatigue, corrosion, or accidental damage occur within the operational life of the aeroplane, the remaining structure can withstand reasonable loads without failure or excessive structural deformation until the damage is detected. Included are the considerations historically associated with fail-safe design. The evaluation should encompass establishing the components which are to be designed as damage-tolerant, defining the loading conditions and extent of damage, conducting sufficient representative tests and/or analyses to substantiate the design objectives (such as life to crack-initiation, crack propagation rate and residual strength) have been achieved and establishing data for inspection programmes to ensure detection of damage. Interpretation of the test results should take into account the scatter in crack propagation rates as well as in lives to crack-initiation. Test results should be corrected to allow for variations between the specimen and the aeroplane component thickness and sizes. This evaluation applies to either single or multiple load path structure.

(b) Identification of Principal Structural Elements. Principal structural elements are those which contribute significantly to carrying flight, ground, and pressurisation loads, and whose failure could result in catastrophic failure of the aeroplane. Typical examples of such elements are as follows:

(1) Wing and empennage: control surfaces, slats, flaps and their attachment hinges and fittings; integrally stiffened plates; primary fittings; principal splices; skin or reinforcement around cutouts or discontinuities; skin-stringer combinations; spar caps; and spar webs.

(2) Fuselage: circumferential frames and adjacent skin; door frames; pilot window posts; pressure bulkheads; skin and any single frame or stiffener element around a cutout; skin or skin splices, or both, under circumferential loads; skin or skin splices, or both, under fore-and-aft loads; skin around a cutout; skin and stiffener combinations under fore-and-aft loads; and window frames.

(c) Inaccessible Areas. Every reasonable effort should be made to ensure inspectability of all structural parts, and to qualify them under the damage-tolerance provisions. In those cases where inaccessible and uninspectable blind areas exist, and suitable damage tolerance cannot practically be provided to allow for extension of damage into detectable areas, the structure should be shown to comply with the fatigue (safe-life) requirements in order to ensure its continued airworthiness. In this respect particular attention should be given to the effects of corrosion.

(d) Testing of Principal Structural Elements. The nature and extent of tests on complete structures or on portions of the primary structure will depend upon applicable previous design, construction, tests, and service experience, in connection with similar structures. Simulated cracks should be as representative as possible of actual fatigue damage. Where it is not practical to produce actual fatigue cracks, damage can be simulated by cuts made with a fine saw, sharp blade, guillotine, or other suitable means. In those cases where bolt failure, or its equivalent, is to be simulated as part of a possible damage configuration in joints or fittings, bolts can be removed to provide that part of the simulation, if this condition would be representative of an actual failure under typical load. Where accelerated crack propagation tests are made, the possibility of creep cracking under real time pressure conditions should be recognised especially as the crack approaches its critical length.

(e) Damage-Tolerance Analysis and Tests. It should be determined by analysis, supported by test evidence, that the structure with the extent of damage established for residual strength evaluation can withstand the specified design limit loads (considered as ultimate loads), and that the damage growth rate under the repeated loads expected in service (between the time at which the damage becomes initially detectable and the time at which the extent of damage reaches the value for residual strength evaluation) provides a practical basis for development of the inspection programme and procedures of this AMC. The repeated loads should be as defined in the loading, temperature, and humidity spectra. The loading conditions should take into account the effects of structural flexibility and rate of loading where they are significant.

(f) Inspection. Detection of damage before it becomes dangerous is the ultimate control in ensuring the damage-tolerance characteristics of the structure. Therefore, the applicant should provide sufficient guidance information to assist operators in establishing the frequency, extent, and methods of inspection of the critical structure, and this kind of information must be included in the maintenance manual. Due to the inherent complex interactions of the many parameters affecting damage tolerance, such as operating practices, environmental effects, load sequence on crack growth, and variations in inspection methods, related operational experience should be taken into account in establishing inspection procedures. It is extremely important to ensure by regular inspection the detection of damage in areas vulnerable to corrosion or accidental damage. However, for crack initiation arising from fatigue alone, the frequency and extent of the inspections may be reduced during the period up to the demonstrated crack-free life of the part of the structure, including appropriate scatter factors.

3.4.3 Fatigue (Safe-Life) Evaluation

The evaluation of structure under the following fatigue (safe-life) strength evaluation methods is intended to ensure that catastrophic fatigue failure, as a result of the repeated loads of variable magnitude expected in service, will be avoided throughout the structure's operational life. Under these methods the fatigue life of the structure should be determined. The evaluation should include the following:

(a) Estimating, or measuring the expected loading spectra for the structure;

(b) Conducting a structural analysis including consideration of the stress concentration effects;

(c) Performing fatigue testing of structure which cannot be related to a test background to establish response to the typical loading spectrum expected in service;

(d) Determining reliable replacement times by interpreting the loading history, variable load analyses, fatigue test data, service experience, and fatigue analysis;

(e) Evaluating the possibility of fatigue initiation from sources such as corrosion, stress corrosion, disbonding, accidental damage and manufacturing defects based on a review of the design, quality control and past service experience; and

(f) Providing necessary maintenance programs and replacement times to the operators. The maintenance program should be included in Instructions for Continued Airworthiness.

3.5 Material Strength Properties and Material Design Values

3.5.1 Definitions

Material strength properties. Material properties that define the strength related characteristics of any given material. Typical examples of material strength properties are: ultimate and yield values for compression, tension, bearing, shear, etc.

Material design values. Material strength properties that have been established based on the requirements or other means as defined in this AMC. These values are generally statistically determined based on enough data that when used for design, the probability of structural failure due to material variability will be minimized. Typical values for moduli can be used.

Aeroplane operating envelope. The operating limitations defined for the product under Subpart G of CS－25.

3.5.2 Statistically Based Design Values

Design values must be based on sufficient testing to assure a high degree of confidence in the values. In all cases, a statistical analysis of the test data must be performed.

The test specimens used for material property certification testing should be made from material produced using production processes. Test specimen design, test methods and testing should:

(ⅰ) conform to universally accepted standards such as those of the American Society for Testing Materials (ASTM), European Aerospace Series Standards (EN), International Standard Organisation (ISO), or other national standards acceptable to the Agency, or:

(ⅱ) conform to those detailed in the applicable chapters/sections of "The Metallic Materials Properties Development and Standardization (MMPDS) handbook", MIL－HDBK－17, ESDU 00932 or other accepted equivalent material data handbooks, or:

(ⅲ) be accomplished in accordance with an approved test plan which includes definition of test specimens and test methods. This provision would be used, for example, when the material design values are to be based on tests that include effects of specific geometry and design features as well as material.

The Agency may approve the use of other material test data after review of test specimen design, test methods, and test procedures that were used to generate the data.

3.5.3 Consideration of Environmental Conditions

The material strength properties of a number of materials, such as non-metallic composites and adhesives, can be significantly affected by temperature as well as moisture absorption. For these materials, the effects of temperature and moisture should be accounted for in the determination and use of material design values. This determination should include the extremes of conditions encountered within the aeroplane operating envelope. For example, the maximum temperature of a control surface may include effects of direct and reflected solar radiation, convection and radiation from a black runway surface and the maximum ambient temperature. Environmental conditions other than those mentioned may also have significant effects on material design values for some materials and should be considered.

Part II Analysis, Study and Exercises

Words

absorption	[əb'zɔːpʃn]	*n*. 吸收；专心致志
accelerometer	[əkˌselə'rɒmɪtə]	*n*. 加速度计
accordance	[ə'kɔːdns]	*n*. 按照，依据，一致
adequate	['ædɪkwət]	*adj*. 充分地；足够地；适当地
adhesive	[ad'hesɪv]	*n*. 黏合剂 *adj*. 黏合的
adjustment	[ə'dʒʌstmənt]	*n*. 调校
aerodynamic	[ˌeərəʊdaɪ'næmɪk]	*adj*. 空气动力学的
airframe	['eəfreɪm]	*n*. 机身
ambient	['æmbɪənt]	*adj*. 周围的；外界的
amended	[ə'mendɪd]	*adj*. 修正的；改进的
anisotropic	[ænˌaɪsəʊ'trɒpɪk]	*adj*. 各向异性的；非均质的
anticipate	[æn'tɪsɪpeɪt]	*v*. 认为……很有可能；期望；预料；先于……行动；盼望；为……早做准备；在……之前来到
applicable	[ə'plɪkəbl]	*adj*. 可适用的；可应用的
arrest	[ə'rest]	*v*. 阻止
assure	[ə'ʃʊə(r)]	*v*. 保证，担保
bar	[bɑː]	*n*. 棒材

bending	[ˈbendɪŋ]	*n*. 弯曲度
bi-directional	[ˌbaɪdɪˈrekʃənəl]	*adj*. 双向的；双指向性的
buffeting	[ˈbʌfɪtɪŋ]	*n*. 振动；猛击；狂暴；肆虐
calibrated	[ˈkælɪbreɪtɪd]	*adj*. 校准过的
catastrophic	[ˌkætəˈstrɒfɪk]	*adj*. 大灾祸的；极不幸的
cavity	[ˈkævətɪ]	*n*. 孔
characteristic	[ˌkærəktəˈrɪstɪk]	*n*. 特性
checklist	[ˈtʃeklɪst]	*n*. 检查表
coalesce	[ˌkəʊəˈles]	*n*. 接合；结合
complaint	[kəmˈpleɪnt]	*n*. 投诉
complement	[ˈkɒmplɪm(ə)nt]	*n*. 补充；补足(处理)
compliance	[kəmˈplaɪəns]	*n*. 服从；可塑性；柔量；承诺；符合性
composite	[ˈkɒmpəzɪt]	*n*. 复合材料；合成物 *adj*. 复合的，合成的
compression	[kəmˈpreʃən]	*n*. 加压；压力；紧缩；压缩，压榨
conceivable	[kənˈsiːvəb(ə)l]	*adj*. 可想到的；可理解的
configuration	[kənˌfɪgəˈreɪʃn]	*n*. 布局；结构；构造；配置；构型
conservative	[kənˈsɜːvətɪv]	*adj*. 保守的
contamination	[kənˌtæmɪˈneɪʃn]	*n*. 污染
convection	[kənˈvekʃn]	*n*. 对流；传送
correlate	[ˈkɒrəleɪt]	*v*. 关联
correlation	[ˌkɒrəˈleɪʃn]	*n*. 联系
corrosion	[kəˈrəʊʒ(ə)n]	*n*. 腐蚀
criteria	[kraɪˈtɪərɪə]	*n*. 标准，条件
cutout	[ˈkʌtaʊt]	*n*. 切口
damping	[ˈdæmpɪŋ]	*n*. 阻尼
deceleration	[ˌdiːseləˈreɪʃn]	*n*. 减速
declaration	[ˌdekləˈreɪʃn]	*n*. 声明
deflection	[dɪˈflekʃn]	*n*. (尤指击中某物后)突然转向，偏斜
deformation	[ˌdiːfɔːˈmeɪʃn]	*n*. 损形；变形；畸形；破相；变丑；残废
demonstrate	[ˈdemənstreɪt]	*v*. 证明；展示；论证；演示
demonstration	[ˌdemənˈstreɪʃn]	*n*. 示范
derivation	[ˌderɪˈveɪʃn]	*n*. 推导
derivative	[dɪˈrɪvətɪv]	*n*. 派生词；衍生字；派生物；衍生物 *adj*. 模仿他人的；缺乏独创性的
designated	[ˈdezɪgneɪtɪd]	*adj*. 指定的
detectability	[dɪˌtektəˈbɪlɪtɪ]	*adj*. 可探测性

deterioration	[dɪˌtɪrɪə'reɪʃn]	*n*. 劣化
detrimental	[ˌdetrɪ'mentl]	*adj*. 有害的；不利的
disbanding	[dɪs'bændɪŋ]	*n*. 脱黏
discontinuity	[dɪsˌkɒntɪ'njuːɪtɪ]	*n*. 不连贯;不连续；间断;中断;悬殊
discrete	[dɪ'skriːt]	*adj*. 离散的
distribution	[ˌdɪstrɪ'bjuːʃn]	*n*. 分布
effectiveness	[ɪ'fektɪvnəs]	*n*. 有效性
empennage	[em'penɪdʒ]	*n*. 尾翼
encompass	[ɪn'kʌmpəs]	*v*. 包含,包括,围绕
encounter	[ɪn'kaʊntə]	*v*. 碰撞
endanger	[ɪn'deɪndʒə]	*v*. 危及
endurance	[ɪn'djʊərəns]	*n*. 耐久性
energy	['enədʒɪ]	*n*. 能量
entail	[ɪn'teɪl]	*v*. 使成为必需;使成为必然
envelope	['envələʊp]	*n*. 范围;信封,封皮
equilibrium	[ˌiːkwɪ'lɪbrɪəm]	*n*. 平衡
equivalant	[ɪ'kwɪvələnt]	*adj*. 相等的
equivalency	[ɪ'kwɪvələnsɪ]	*n*. 等效性
extrapolate	[ɪk'stræpəleɪt]	*v*. 推断，推算
extrapolation	[ɪkˌstræpə'leɪʃn]	*n*. 推断；外推;外插
extrusion	[ɪk'struːʒn]	*n*. 型材；挤压
fabrication	[ˌfæbrɪ'keɪʃn]	*n*. 制造
fatigue	[fə'tiːg]	*n*. 疲劳,疲乏
		v. 使疲劳
finite	['faɪnaɪt]	*adj*. 有限的
fitting	['fɪtɪŋ]	*n*. 设备；连接件
formula	['fɔːmjʊlə]	*n*. 公式；方程式；计算式；分子式；方案；方法
fuselage	['fjuːzəlɑːʒ]	*n*. 机身
gauge	[geɪdʒ]	*n*. 量具;量器;厚度;尺寸
geometry	[dʒɪ'ɒmətrɪ]	*n*. 几何学
gust	[gʌst]	*n*. 爆发;一阵突发
hinge	[hɪn(d)ʒ]	*n*. 铰链;枢纽;关键;铰合部
identical	[aɪ'dentɪkl]	*adj*. 完全相同的；相同的；同一的
impedance	[ɪm'piːdns]	*n*. 阻抗
inaccessible	[ˌɪnək'sesɪb(ə)l]	*adj*. 难以接近的
inertia	[ɪ'nɜːʃə]	*n*. 惯性

ingestion	[ɪnˈdʒestʃən]	*n*. 吸收;吞下
inservice	[ɪnˈsɜːvɪs]	*adj*. 服役中的
inspectability	[ɪnˈspektɪbɪlɪtɪ]	*n*. 可检验性
intensity	[ɪnˈtensətɪ]	*n*. 强度
interpretation	[ɪnˌtɜːprəˈteɪʃn]	*n*. 理解;解释;说明
jeopardise	[ˈdʒepədaɪz]	*v*. 危及
joint	[dʒɒɪnt]	*n*. (尤指构成角的)接头; 关节; 连接开口
judgement	[ˈdʒʌdʒmənt]	*n*. 判断
justification	[ˌdʒʌstɪfɪˈkeɪʃn]	*n*. 正当理由
likelihood	[ˈlaɪklɪhʊd]	*n*. 可能性
longitudinal	[ˌlɒŋgɪˈtjuːdɪn(ə)l]	*adj*. 纵向的
malfunction	[ˌmælˈfʌŋkʃn]	*n*. 故障,失灵, 机能失常
maneuver	[məˈnuːvə(r)]	*n*. 操控; 特技动作;熟练动作; 机动动作
margin	[ˈmɑːdʒən]	*n*. 幅度,差额
metallic	[məˈtælɪk]	*adj*. 金属的,含金属的
methodology	[ˌmeθəˈdɒlədʒɪ]	*n*. 方法论
minimise	[ˈmɪnəˌmaɪz]	*v*. 最小化,最少化
modulus	[ˈmɔːdjələs]	*n*. 系数,模数
moisture	[ˈmɒɪstʃə]	*n*. 水分;湿度
nature	[ˈneɪtʃə]	*n*. 性质
non-deterministic	[nʌndɪˌtɜːmɪˈnɪstɪk]	*adj*. 不确定的
objective	[əbˈdʒektɪv]	*n*. 目标
occupant	[ˈɒkjəpənt]	*n*. 使用者,居住者, 乘员
pane	[peɪn]	*n*. 板
partial	[ˈpɑːʃ(ə)l]	*adj*. 部分的;不完全的
penetration	[ˌpenəˈtreɪʃn]	*n*. 渗透
permanent	[ˈpɜːmənənt]	*adj*. 永久的;永恒的;长久的
pertinent	[ˈpəːtɪnənt]	*adj*. 切题的,相关的
pivot	[ˈpɪvət]	*n*. 中心点;基准部队 *v*. 绕……旋转;以……为轴转动
plate	[pleɪt]	*n*. 板材
ply	[plaɪ]	*n*. 层
porosity	[pɔːˈrɒsətɪ]	*n*. 孔隙度
portion	[ˈpɔːʃ(ə)n]	*n*. 部分
preclude	[prɪˈkluːd]	*v*. 排除
predict	[prɪˈdɪkt]	*n*. 预言; 预告; 预报; 预测

premium	[ˈpriːmɪəm]	*adj*. 高级的;优质的
pressurization	[ˌpreʃərɪˈzeɪʃn]	*n*. 增压;气密;压力输送; 加压
propagation	[ˌprɒpəˈgeɪʃn]	*n*. 扩展
property	[ˈprɒpətɪ]	*n*. 性能,性质;财产; 特性
provision	[prəˈvɪʒn]	*n*. 条款, 规定
quantitative	[ˈkwɒntɪtətɪv]	*adj*. 用数量表示的;定量的
radiation	[ˌreɪdɪˈeɪʃn]	*n*. 辐射;放射物
radiographic (X-ray)	[ˌreɪdɪəʊˈgræfɪk]	*n*. 射线照相(X 射线)
reinforcement	[ˌriːɪnˈfɔːsm(ə)nt]	*n*. 加强; 强化; 钢筋
representativeness	[ˌreprɪˈzentətɪvnɪs]	*n*. 代表性
residual	[rɪˈzɪdjʊəl]	*adj*. 残余的
retroactive	[ˌretrəʊˈæktɪv]	*adj*. 追溯的;有追溯效力的;反动的
scenario	[səˈnɑːrɪəʊ]	*n*. 方案, 情况
scope	[skəʊp]	*n*. 范围
sequence	[ˈsiːkw(ə)ns]	*n*. 顺序;连续事件
shear	[ʃɪə(r)]	*v*. 切变,剪,断裂
shielding	[ˈʃiːldɪŋ]	*n*. 屏蔽
shrinkage	[ˈʃrɪŋkɪdʒ]	*n*. 收缩
sideslip	[ˈsaɪdslɪp]	*n*. 侧滑, 横滑
simultaneous	[ˌsɪm(ə)lˈteɪnɪəs]	*adj*. 同时的
solidify	[səˈlɪdɪfaɪ]	*v*. 凝固
sophisticated	[səˈfɪstɪkeɪtɪd]	*adj*. 复杂的
spar	[spɑː]	*n*. 圆材;翼梁
specification	[ˌspesɪfɪˈkeɪʃn]	*n*. 规格;说明书;详述
specify	[ˈspesɪfaɪ]	*v*. 具体说明; 明确规定; 详述; 详列
specimen	[ˈspesɪmən]	*n*. 样品; 样本; 标本; (尤指动植物的)单一实例; (化验的)抽样,血样,尿样
spectrum	[ˈspektrəm]	*n*. 谱;光谱;范围
splice	[ˈsplaɪs]	*n*. 胶接处; 粘接处; 铰接处
stabilizer	[ˈsteɪbɪlaɪzə]	*n*. 稳定器; (飞机的)安定面
standardization	[ˌstændədaɪˈzeɪʃ(ə)n]	*n*. 标准化;规格化
statistically	[stəˈtɪstɪklɪ]	*adv*. 统计地,统计学上地
stiffness	[ˈstɪfnəs]	*n*. 硬度, 刚度
stipulate	[ˈstɪpjʊleɪt]	*n*. 规定
stringer	[ˈstrɪŋə]	*n*. 特约记者;桁条
sub-component	[sʌb kəmˈpəʊnənt]	*n*. 子元素
substantiate	[səbˈstænʃɪeɪt]	*v*. 证实;证明

survey	[sə'veɪ]	*v*. 审视;测量;勘测;测绘;调查 *n*. 概论;全面的考察;概况
taxiing	['tæksɪɪŋ]	*n*. 滑行
tension	['tenʃn]	*n*. 张力,拉力;紧张
tolerance	['tɒlərəns]	*n*. 耐受性,容限,容忍
torque	[tɔːk]	*n*. 扭矩
transient	['trænʃnt]	*adj*. 瞬时的
turbulence	['təːbjʊl(ə)ns]	*n*. 湍流
ultimate	['ʌltɪmət]	*adj*. 最大的;极端的
uncertainty	[ʌn'sɜːtntɪ]	*n*. 不确定性
underpredict	['ʌndə prɪ'dɪkt]	*v*. 低估
unsymmetric	[ˌʌnsɪ'metrɪk]	*adj*. 不对称的
utilisation	[ˌjuːtɪlaɪ'zeɪʃən]	*n*. 利用
valid	['vælɪd]	*adj*. 有效的
validate	['vælɪdeɪt]	*v*. 证实; 确认; 使有法律效力; 批准; 确认……有效; 认可;验证;使合法有效;宣布合法有效
validation	[ˌvælə'deɪʃən]	*n*. 生效;批准;验证;确认
value	['væljuː]	*n*. 值
variability	[ˌveərɪə'bɪlətɪ]	*n*. 可变性,变化性
verify	['verɪfaɪ]	*v*. 核实; 查对; 核准; 证明; 证实
weld	[weld]	*v*. 焊接
withstand	[wɪð'stænd]	*v*. 经受,承受;顶住,抵住

Phrases & Expressions

accurate representation	精确表示
additional precaution	额外的预防措施
adequately predicted	适当地预测
adjacent supporting airframe structure	邻近的机身支持结构
adverse condition	不利条件
aeroelastic stability	气动弹性稳定性
alerting system	警戒系统
ambient temperature	环境温度
assessment criteria	评估标准
automatically or by operational procedure	自动或操作程序
bird strike	鸟撞
blade failure	叶片失效

boundary condition	边界条件
bracket attachment	支架附件；托架连接件
braking system	刹车系统
cabin pressure differential load	客舱压差载荷
case by case basis	具体情况具体分析
cast part	铸件
certification specification	审定规范
coefficients of variation	变异系数
complete package of elements	完整的一系列元素
compliance finding	符合性结果
confirmatory test	验证性测试
consequence of failure	失败后果
conservative calculated load	保守计算的载荷
consistent mechanical properties	一致的力学性能
continued applicability	持续适用性
control system flexibility	控制系统柔性
controlled cooling	控制冷却
cope with the situation	来应对这种情况
correction factor	校正因子
corrective action	纠正措施
crack initiation	裂纹开裂；裂纹萌生,起裂，初始裂纹
crack propagation rate	裂纹扩展速率
crack stopper	止裂器；止裂带
creep cracking	蠕变断裂
damage tolerance	损伤容限；耐损伤性
derive from	来源于
design value	设计值
detectable size	可检尺寸
detrimental permanent deformation	有害的永久性变形
dynamic characteristics	动态特性
dynamic effect	动态效应
elevated stress level	高应力水平
enhanced safety	提高安全性
environmental degradation	环境退化
exceptional pilot skill	高超的飞行员技能
excite various portions of the structure	激发结构的各个部分
external visibility	外部能见度

extremely improbable	极不可能
fatigue damage arrested	疲劳损伤被阻止
feasible weight	可行重量
flight envelope	飞行包线
flight mechanics	飞行力学
forged product	锻造产品
fretting condition	微动条件；微振磨损情况
functional safety standard	功能安全标准
gear load	齿轮(传动)载荷
ground gust	地面阵风
high dielectric strength	高绝缘强度
in lieu of	替代
inaccessible area	难以检测区域
inadvertent action	疏忽行为
input data	输入数据
internal linkage	内在联系
landing gear in the extended, retracting and retracted positions	起落架在放下、收起时和收起后的位置
landing impact	着陆撞击；着陆冲击
leading edge devices	前缘装置
lightning discharge	雷电放电
lightning diverters	避雷器
lightning strike	雷击
liquid penetrant techniques	液体渗透剂技术
loading history	载荷历程；载荷履历
loading spectrum	载荷谱
magnetic particle inspection	磁粉探伤
margins of safety	安全边界(安全的极限)
material variability	材料可变性
maximum working stress level	最大工作应力水平
mechanical robustness	机器鲁棒性
metallographic sample	金相试样
microstructural characteristics	微观结构特征
moisture absorption	吸湿性
monitoring of the process	过程监控
nearly identical	几乎相同
negative effect	消极影响；负面影响

nondestructive inspection	无损检验
nonopenable window	不可打开的窗户
normal tolerances	标准公差
nose-wheel steering	前轮转向
notch sensitive	缺口敏感
operating envelope	工作包线
operating limitations	操作限制
overall configuration	整体配置
overall integrity	整体的完整性
parametric variation	参数变化
partial failure	局部失效；局部故障
pitch maneuver	俯仰操作
potential risk	潜在风险
potentially hazardous operating condition	潜在危险操作条件
pressure belt	测压带
pressurised cabin	增压舱室
primary aerodynamic surface	主要气动表面
primary conductor	主要导体
protective coating	防护涂层
qualification of the process	工艺鉴定
quantitative relationship	定量关系
radiographic inspection	射线探测
rapid expansion	快速膨胀
reasonable pilot technique	合理的飞行员技术
residual internal stress	残余内应力
residual static strength	剩余静强度
residual strength	剩余强度
reverse thrust	反推力
revised analysis	修正分析；改进的分析
rigid body mode	刚体模态
roll maneuvers	横滚操作
runway surface roughness	跑道表面粗糙度
scatter factor	离散因子
serious consequence	严重的后果
set forth	(前文)阐述过、规定的内容；陈述
shock absorption	减震
simultaneous failure	同时失效；同时发生的裂纹

softening temperature	软化温度
spacing of structural members	结构构件的间距；结构件间隔
spectrum severity	频谱严重度；(载荷)谱的严重性
standardised inspection	标准化检验
statistically determined	统计确定的
strain gauge	应变计
strength properties	强度特性
stress concentration	应力集中
stress corrosion	应力腐蚀
structural damping	结构阻尼
structural integrity	结构完整性
structural representation	结构代表性
suitable temperature	合适的温度
system deactivation	系统停用
take-off delayed	起飞延误
taxiway slope	滑行道坡度
temperature effects	温度影响；温度效应
tensile ultimate strength	拉伸极限强度
tensile yield strength	拉伸屈服强度
test bar	试棒
test coupon	试样
test evidence	测试佐证
test specimen	试样
threat model	威胁模型
three-dimensional structure	三维结构
total loss	完全丧失
toughened glass	钢化玻璃
towing vehicle	拖车
tyre burst	轮胎爆裂
tyre characteristics	轮胎特性
tyre debris	轮胎碎片
uninspectable blind area	不可检盲区
asymmetric load	不对称载荷
unusual design feature	不寻常的设计特性
unwanted warning	不必要的警告
variability in mechanical properties	机械性能变化
variable load	变载荷

vary accordingly 相应变化
warning system 告警系统
wind tunnel test 风洞试验
yaw maneuver 偏航操作

Abbreviations

AMC Acceptable Means of Compliance 可接受的执行方法

Sentence Comprehension

1. This AMC sets forth an acceptable means, but not the only means, of demonstrating compliance with the provisions of CS-25 related to the validation, by flight load measurements, of the methods used for determination of flight load intensities and distributions, for large aeroplanes.

   译文：这份 AMC 规定了一种可接受的方法，即通过飞行载荷测量验证大型飞机飞行载荷强度和分布的方法，以证明它们符合 CS-25 的规定，但这不是唯一的方法。

2. The scope of this AMC however is limited to validation of methods used for determination of loads intensities and distributions by flight load measurements.

   译文：然而，AMC 的范围仅限于验证通过飞行载荷测量来确定载荷强度和分布的方法。

3. Products requiring a new type certificate will in general require flight-test validation of flight loads methods unless the Applicant can demonstrate to the Agency that this is unnecessary.

   译文：需要型号合格证的产品一般需要用飞行试验来验证飞行载荷方法，除非申请人能向局方证明这是不必要的。

4. Applicants who are making a change to a Type Certificated airplane, but who do not have access to the Type Certification flight loads substantiation for that airplane, will be required to develop flight loads analyses, as necessary, to substantiate the change.

   译文：如果申请人要更改一架已经过型号认证的飞机，但无法获得该飞机的型号合格飞行载荷证明，则需要根据需要进行飞行载荷分析，以证实更改(的适用性)。

5. The uncertainties in both the flight testing measurements and subsequent correlation should be carefully considered and compared with the inherent assumptions and capabilities of the process used in analytic derivation of flight loads.

   译文：飞行试验测量和随后的相关性分析中的不确定性应该被仔细考虑，并与用于分析推导飞行载荷的过程的固有假设和能力进行比较。

6. The application of methods such as Finite Element Method or engineering formulas to complex structures in modern aircraft is considered reliable only when validated by full scale tests (ground and/or flight tests).

 译文: 有限元方法或工程公式(计算)等方法在现代飞机复杂结构中的应用,只有经过全尺寸试验(地面和/或飞行试验)的验证才被认为是可靠的。

7. Sufficient limit load test conditions should be performed to verify that the structure meets the deformation requirements of CS 25. 305(a) and to provide validation of internal load distribution and analysis predictions for all critical loading conditions.

 译文: 应执行充分的极限荷载试验条件,以验证结构满足 CS 25. 305(a)的变形要求,并对所有临界荷载条件下的内部荷载分布和分析预测进行验证。

8. CS 25. 362 ("Engine failure loads") requires that the engine mounts, pylons, and adjacent supporting airframe structure be designed to withstand 1 g flight loads combined with the transient dynamic loads resulting from each engine structural failure condition.

 译文: CS 25. 362("发动机失效载荷")要求发动机支架、挂架和相邻的支撑机体结构在设计时要能承受 1g 的飞行载荷和由发动机结构失效引起的瞬态动态载荷。

9. The objective of the analysis methodology is to develop acceptable analytical tools for conducting investigations of dynamic engine structural failure events.

 译文: 分析方法的目标是开发可接受的分析工具,进行发动机结构动态故障事件的调查。

10. The purpose of the requirement was to protect the flight control system from excessive peak ground wind loads while the aeroplane is parked or while taxiing downwind.

 译文: 该要求的目的是保护飞行控制系统在飞机停放或顺风滑行时免受过量的地面峰值风载荷。

11. The ground gust requirements take into account the conditions of the aeroplane parked with controls locked, and taxiing with controls either locked or unlocked.

 译文: 地面阵风要求考虑了飞机在控制锁定条件下停放,和在控制锁定或解锁条件下滑行几种情况。

12. Guidance information is provided for showing compliance with CS 25. 491, relating to structural design for aeroplane operation on paved runways and taxiways normally used in commercial operations.

 译文: 提供指导材料用以表明符合 CS25. 491 条款。该条款涉及飞机在有铺面的跑道上和在通常用于商业运作的滑行道上运行时的结构要求。

13. Because the various methods described above produce different results, the guidance information given in paragraphs 4, 5, and 6 of this AMC should be used when demonstrating compliance with CS 25.491.

译文：因为上面描述的各种方法会产生不同的结果，所以在证明符合 CS 25.491 时，应该使用 AMC 的第 4、5 和 6 段中给出的指导信息。

14. The loading spectrum should be based on measured statistical data of the type derived from government and industry load history studies and, where insufficient data are available, on a conservative estimate of the anticipated use of the aeroplane.

译文：载荷谱应基于政府和行业载荷历史研究中获得同类飞机的实测统计数据，如果数据不充分，则应基于对飞机预期用途的保守估计。

15. The damage-tolerance evaluation of structure is intended to ensure that should serious fatigue, corrosion, or accidental damage occur within the operational life of the aeroplane, the remaining structure can withstand reasonable loads without failure or excessive structural deformation until the damage is detected.

译文：结构的耐损伤性评估旨在确保在飞机的使用寿命内出现严重的疲劳、腐蚀或意外损伤时，其余结构可以承受合理的载荷而不会发生故障或结构过度变形，直到检测到损伤为止。

16. In practice it may not be possible to guard against the effects of multiple damage and fail-safe substantiation may be valid only up to a particular life which would preclude multiple damage.

译文：在实践中，可能无法预防多重损伤的影响，并且失效安全证明仅在一个可以避免多重损伤的特定寿命内才有效。

17. Detection of damage before it becomes dangerous is the ultimate control in ensuring the damage-tolerance characteristics of the structure.

译文：在结构发生危险前进行损伤检测是保证结构损伤容限特性的最终控制方法。

18. It is extremely important to ensure by regular inspection the detection of damage in areas vulnerable to corrosion or accidental damage.

译文：通过定期检查来确保在易受腐蚀或意外损坏的地方发现损坏是极为重要的。

19. The test article should be full scale (component or sub-component) and represent that portion of the production aircraft requiring test.

译文：测试物品应是全尺寸的(部件或子部件)，并代表生产飞机需要测试的那部分。

20. All differences between the test article and production article should be accounted for either by analysis supported by test evidence or by testing itself.

    译文：测试项目和生产项目之间的所有差异都应该由测试证据支持的分析或测试本身来解释。

21. Confirmatory tests may be required to check the adequacy of the lightning protection provided (e. g. , to confirm the adequacy of the location and size of bonding strips on a large radome).

    译文：可能需要进行验证性测试来检查所提供的雷电防护是否足够(例如，确认大型雷达天线罩上的连接带位置和大小是否足够)。

22. These values are generally statistically determined based on enough data that when used for design, the probability of structural failure due to material variability will be minimized.

    译文：这些数值通常是根据足够的数据统计确定的，当用于设计时，由于材料变异性引起结构失效的概率是最小的。

23. The test specimens used for material property certification testing should be made from material produced using production processes.

    译文：用于材料性能认证试验的试样应由使用生产工艺生产的材料制成。

24. Casting is a method of forming an object by pouring molten metal into a mould, allowing the material to solidify inside the mould, and removing it when solidification is complete.

    译文：铸造是将熔融的金属倒入模具中，使材料在模具内凝固，待凝固完成后取出的一种成型方法。

25. The majority of these characteristics can be detected, evaluated, and quantified by standard nondestructive testing methods, or from destructive methods on prolongation or casting cut-up tests.

    译文：这些特性中的大多数可以通过标准的无损检测方法，或通过拉伸试验或压铸试验中的破坏性方法来检测、评价和定量。

26. The casting process is capable of producing a casting with uniform properties throughout the casting or, if not uniform, with a distribution of material properties that can be predicted to an acceptable level of accuracy.

    译文：该铸造工艺能够在整个铸造过程中生产出具有均匀性能的铸件，如果不是均匀的，也可以将材料性能的分布预测到可接受的精度水平。

27. Guidance on non-destructive inspection techniques and methods can be obtained from national and international standards.

    译文：对无损检测技术和方法的指导可以从国家和国际标准中获得。

28. When the process has been qualified, it should not be altered without completing comparability studies and necessary testing of differences.

译文：当这个过程已经被鉴定是合格的时，则在未完成可比性研究和必要的差异试验之前不得改变它。

29. Deviations from the limits established in the process qualification and product proving programs should be investigated and corrective action taken.

译文：应对偏离于工艺鉴定和产品验证程序中建立起来的门限值的情况进行调查，并采取纠正措施。

30. A number of aeroplane accidents have occurred because the aeroplane was not properly configured for take-off and a warning was not provided to the flight crew by the take-off configuration warning system.

译文：许多飞机事故的发生，是因为飞机没有正确地配置起飞，起飞构型警告系统没有向机组人员提供警告。

31. It has been Aviation Authorities policy to categorise systems designed to alert the flight crew of potentially hazardous operating conditions as being at a level of criticality associated with a probable failure condition.

译文：航空当局的政策是，将旨在提醒机组人员潜在危险操作条件的系统归类为与可能的故障条件相关的临界级别。

32. The EASA Agency approved Master Minimum Equipment List (MMEL) includes those items of equipment related to airworthiness and operating regulations and other items of equipment which the Agency finds may be inoperative and yet maintain an acceptable level of safety by appropriate conditions and limitations.

译文：欧洲航空安全局(EASA)批准的主最低设备清单(MMEL)包括与适航性和操作规程相关的设备项目，以及该机构认为可能失效但通过适当的条件和限制保持可接受的安全水平的其他设备项目。

33. During flight testing it should be shown that the take-off configuration warning system does not issue nuisance alerts or interfere with other systems.

译文：在飞行测试期间，应该显示起飞构型警报系统不会发出干扰警报或干扰其他系统。

34. The design take-off weight and the design landing weight conditions should both be included as configurations subjected to energy absorption tests.

译文：设计起飞重量和设计着陆重量条件都应该包括在经过能量吸收测试的配置中。

35. In-service experience shows that traditional large transport aeroplane configurations, featuring high aspect ratio wings built around a single torsion box manufactured of light metal alloy, have demonstrated inherent structural robustness with regard to wheel and tyre debris threats.

译文：在役经验表明，传统的大型运输飞机配置，具有建立在一个由轻金属合金制造的单一扭力盒四周的高展弦比机翼，显示出固有的结构坚固性，以应对轮胎碎片的威胁。

36. When protection is afforded by the flight crew alerting system, the damage detection means should be independent of the availability of aeroplane power supplies and should be active during ground maneuvering operations effected by means independent of the aeroplane.

译文：当机组人员警报系统提供保护时，损害检测手段应独立于飞机电源供应，并应在使用独立于飞机的手段进行地面操纵操作时发挥作用。

37. The systems should be designed so that no malfunction or failure of one system will adversely affect the other.

译文：系统的设计应使一个系统的功能不正常或失效不会对另一个系统产生不利影响。

38. It should be shown by flight tests that exceptional pilot skill is not required to land the aeroplane using the normal aeroplane instruments and the view provided through the main or side windows having the degree of impairment to vision resulting from the encounter of severe hail, birds or insects.

译文：飞行试验应表明，使用普通飞机仪表以及由于遭遇严重冰雹、鸟类或昆虫的影响导致主窗或侧窗提供的视野受到影响的情况下，降落飞机不需要飞行员具有特殊的技能。

39. Typical designs of windshields and cockpit side windows are laminated multi-plied constructions, consisting of at least two structural plies, facing plies, adhesive interlayers, protective coatings, embedded electro-conductive heater films or wires, and mounting structure.

译文：典型的挡风玻璃和驾驶舱侧窗的设计是多层叠层结构，包括至少两个结构层、外面层、黏合夹层、防护涂层、嵌入式导电加热膜或导线以及安装结构。

40. When substantiation is shown by test evidence, the test apparatus should closely simulate the structural behaviour (e. g., deformation under pressure loads) of the aircraft mounting structure up to the ultimate load conditions.

译文：当有试验证据证实时，试验装置应接近模拟飞机安装结构的结构行为（例如，压力载荷下的变形），直至极限载荷条件。

41. The demonstration should account for material characteristics and variability in service material degradation, critical temperature effects, maximum cabin differential pressure, and critical external aerodynamic pressure.

译文：演示应考虑材料特性和使用中材料退化的变异性、临界温度效应、最大客舱压差和临界外部空气动力压力。

42. Inconsistent or inaccurate use of terms may lead to the installation of doors and hatches that do not fully meet the safety objectives of the regulations.

译文: 不一致或不准确的使用条款可能导致安装的门和舱口不能完全满足本条例的安全目标。

43. Service history has shown that to prevent doors from becoming a hazard by opening in flight, it is necessary to provide multiple layers of protection against failures, malfunctions, and human error.

译文: 使用历史表明,为了防止在飞行中打开舱门造成危险,必须提供多层保护,以防止失效、功能不正常和人为错误。

44. Systems used to protect the structure against environmental degradation can have a negative effect on fatigue life and therefore should be included as part of the test article.

译文: 用于保护结构免受环境退化的系统会对疲劳寿命产生负面影响,因此应作为试验条款的一部分。

45. Test load spectrum should be derived based on a spectrum sensitive analysis accounting for variations in both utilisation.

译文: 测试载荷谱应基于频谱敏感性分析得出,考虑两种利用情况的变化。

46. The maximum temperature of a control surface may include effects of direct and reflected solar radiation, convection and radiation from a blank runway surface and the maximum ambient temperature.

译文: 受控表面的最高温度可受直接或反射的太阳辐射,对流和来自空白跑道表面的辐射以及最高环境温度的影响。

47. Determining reliable replacement times by interpreting the loading history, variable load analyses, fatigue test data, service experience.

译文: 通过解释载荷历程、变载荷分析、疲劳试验数据、使用经验来确定可靠的更换时间。

48. In the case where fatigue damage is arrested at a readily detectable size following rapid crack growth or a sudden load path failure under the application of high loads, the structure must be able to withstand the loads defined in CS 25. 571 (b)(1) to (6) inclusive up to that size of damage.

译文: 在高荷载作用下,随着裂纹的快速扩展或突然的荷载路径失效,疲劳损伤被抑制在一个容易检测到的尺寸,结构必须能够承受 CS 25. 571(b)(1)~(6)中定义的荷载,包括该尺寸的损伤。

49. In a pressurised fuselage, an obvious partial failure might be detectable through the inability of the cabin to maintain operating pressure or controlled decompression after occurrence of the damage.

译文: 在增压机身中,通过机舱在损坏发生后无法保持操作压力或控制减压,可以检测到明显的部分故障。

Advanced Word Study

assemble [ə'sembl] *v.* 聚集;集合;收集;装配;组装

Definitions:

① to meet together, to bring together (as in a particular place or for a particular purpose), to come together as a group

② to bring people or things together as a group

③ to fit together all the separate parts of something

Synonyms: cluster, collect, concentrate, conglomerate, convene, converge, gather, meet

Antonyms: break up, disband, disperse, split (up)

Usage examples:

① All the students were asked to *assemble* in the main hall.

② He tried to *assemble* his thoughts.

③ The shelves are easy to *assemble.*

④ An applicant for an original, special airworthiness certificate for a new aircraft, manufactured under a production certificate, including aircraft *assembled* by another person from a kit, is entitled to a special airworthiness certificate without further showing except that the Administrator may inspect the aircraft to determine conformity to the type design and condition for safe operation.

conservative [kən'sɜːvətɪv] *adj.* 保守的

Definitions:

① opposed to great or sudden social change

② showing that you prefer traditional styles and values

③ (of an estimate) lower than what is probably the real amount or number

Synonyms: old-fashioned, old-school, traditional, unprogressive, careful, cautious, considerate

Antonyms: broad-minded, open-minded, nonconventional, nontraditional, progressive, careless, incautious, unmindful

Usage examples: Conservative substantiation of the strength and deformation requirements up to limit and ultimate loads normally requires testing of sub-components, full scale components or full scale tests of assembled components.

derive [dɪ'raɪv] *v.* 获得;取得;得到;起源;产生

Definitions:

① to take, receive, or obtain especially from a specified source

② to trace the derivation of
③ to have or take origin: come as a derivative

Synonyms: conclude, decide, deduce, extrapolate, gather, infer, judge, make out, reason, understand

Usage examples:

① Petroleum is *derived* from coal tar.
② It was *derived* from their observations.
③ Test load spectrum should be *derived* based on a spectrum sensitive analysis.
④ The loading spectrum should be based on measured statistical data of the type *derived* from government and industry load history.

detectable [dɪˈtektəbl] *adj*. 可检测的;可探测的

Definitions:

① able to be perceived by a sense or by the mind
② (especially of something that is not easy to see, hear, etc.) that can be discovered or noticed

Synonyms: appreciable, perceptible, sensible, distinguishable

Antonyms: undetectable, insensible, inappreciable, indistinguishable

Usage Exmaples:

① The noise is barely *detectable* by the human ear.
② In the case where fatigue damage is arrested at a readily *detectable* size following rapid crack growth or a sudden load path failure under the application of high loads, the structure must be able to withstand the loads defined in CS 25. 571(b) (1) to (6) inclusive up to that size of damage.
③ In a pressurised fuselage, an obvious partial failure might be *detectable* through the inability of the cabin to maintain operating pressure or controlled decompression after occurrence of the damage.
④ In any damage determination, including those involving multiple cracks, it is possible to establish the extent of damage in terms of detectability with the inspection techniques to be used, the associated initially *detectable* crack size, the residual strength capabilities of the structure, and the likely damage.

entail [ɪnˈteɪl] *v*. 使成为必需;使成为必然

Definitions:

① to impose, involve, or imply as a necessary accompaniment or result
② to confer, assign, or transmit (something) for an indefinitely long time
③ to fix (a person) permanently in some condition or status

Synonyms: carry, encompass, include, involve

Antonyms: exclude, leave (out), miss out

Usage Exmaples:

① The project will *entail* considerable expense.

② Recorded load and stress data *entails* instrumenting aeroplanes in service to obtain a representative sampling of actual loads and stress experienced.

③ Damage-tolerance design is required, unless it *entails* such complications that an effective damage-tolerant structure cannot be achieved within the limitations of geometry.

margin　　[ˈmɑːdʒən]　　*n.* 幅度,差额,差数

Definitions:

① the empty space at the side of a written or printed page

② the outside limit and adjoining surface of something

③ a spare amount or measure or degree allowed or given for special situations

④ a bare minimum below which or an extreme limit beyond which something becomes impossible or is no longer desirable

⑤ the limit below which economic activity cannot be continued under normal conditions

⑥ measure or degree of difference

⑦ the amount of time, or number of votes, etc. by which somebody wins something

Synonyms: border, boundary, circumference, edge, end, frame, perimeter, periphery

Usage examples:

① The house is situated at the *margin* of the woods.

② The boss left no *margin* for error.

③ Please write your name in the left *margin* of the page.

④ This is usually based on the number of test specimens available, the analytical static strength *margins* of safety of the structure and the range of supporting details or sub-component tests.

reinforcement　　[ˌriːɪnˈfɔːsm(ə)nt]　　*n.* 加强；强化；钢筋；援军

Definitions:

① the action of strengthening or encouraging something

② the state of being reinforced

③ something that strengthens or encourages something

④ an addition of troops, supplies, etc., that augments the strength of an army or other military force

⑤ something designed to provide additional strength (as in a weak area)

⑥ the act of making something stronger, especially a feeling or an idea

Synonyms: brace, support

Antonyms: collapse, looseness, shake

Usage examples:

① We need to prevent enemy *reinforcements* from reaching the front line.

② The bridge is in need of *reinforcement.*

③ The children respond well to praise and positive *reinforcement.*

④ Skin or structural *reinforcement* around cutouts or discontinuities.

simulate [ˈsɪmjʊleɪt] *v.* 假装;冒充;装作;模拟;模仿

Definitions:

① simulate something to pretend that you have a particular feeling

② simulate something to create particular conditions that exist in real life using computers, models, etc., usually for study or training purposes

Synonyms: imitate, reproduce, replicate, duplicate, mimic, parallel, pretend

Usage examples:

① I tried to *simulate* surprise at the news.

② Computer software can be used to *simulate* operating conditions.

③ The process of actually locating where damage should be *simulated* in principal structural elements is identified in paragraph 2.2 of this AMC.

④ The tests should closely *simulate* service loading conditions.

validation [ˌvæləˈdeɪʃən] *n.* 生效;批准;验证;确认

Definitions:

① the act of proving that something is true or correct

② the act of making something legally valid

③ the act of stating officially that something is useful and of an acceptable standard

④ the feeling of recognition from others that you are right or good enough

Synonyms: attestation, confirmation, documentation, evidence, proof, substantiation, witness

Antonyms: disproof

Usage examples:

① There must be some form of external *validation* of the data.

② An agreement on the *validation* of the institution's degree courses.

③ We seek *validation* from our peers.

④ The scope of this AMC however is limited to *validation* of methods used for determination of loads intensities and distributions by flight load measurements.

Exercises for Self-Study

Ⅰ. Translate Sentences

1. Substantiation of the strength and deformation requirements up to limit and ultimate loads normally requires testing of sub-components, full scale components or full scale tests of assembled components (such as a nearly complete airframe).
2. If the test results do not correlate with the analysis, the reasons should be identified and appropriate action taken.
3. For each specific engine design, the applicant should consider whether these types of failures are applicable, and whether they present a more critical load condition than blade loss.
4. Aeroplane design loads should be developed for the most critical conditions arising from taxi, take-off, and landing run.
5. Damage location and growth data should also be considered in establishing a recommended inspection programme.
6. The nature and extent of tests on complete structures or on portions of the primary structure will depend upon applicable previous design and structural tests, and service experience with similar structures.
7. Service and test experience of similar inservice components that were designed using similar design criteria and methods should demonstrate that the load paths and potential failure modes of the components are well understood.
8. With the establishment of consistent production, it is possible to reduce the inspection frequency of the non-visual inspections required by the rule for non-critical castings, with the acceptance of the Agency.
9. If the distribution of mechanical properties derived from these tests is acceptable, when compared to the property values determined in the qualification program, the frequency of testing may be reduced.
10. The panel should be subjected to the most adverse combinations of pressure loading, including the maximum internal pressure, external aerodynamic pressure, temperature effects, and where appropriate, flight loads.

Ⅱ. Multiple Choice Questions

1. The aim is to demonstrate that the complete package delivers reliable or ________ calculated loads.

   A. correct B. conservative C. radical D. neutral

2. The sizing of the structure of the aircraft generally involves a number of steps and requires detailed knowledge of air loads, ________, ________, ________,

flight control system characteristics, etc.

A. stiffness B. damping C. mass D. colour

3. A major section of the airframe structure is which can be tested as a ________ unit to qualify the structure.

A. partial B. total C. complete D. full

4. The evaluation should include the following: ... determining reliable replacement times by interpreting the ________.

A. loading history B. variable load analyses

C. fatigue test data D. service experience

E. fatigue analysis

5. These values are generally statistically determined based on enough data that when used for ________, the probability of structural failure due to material variability will be minimized.

A. manufacture B. design C. production D. operation

6. Materials should be produced using production ________ and processes accepted by the Agency.

A. quantity B. specifications C. requirements D. tools

7. Generally speaking, the article which is ________ is much more accurate.

A. only by imagination B. recently developed

C. statistically determined D. published

8. Multiple load path construction and the use of crack stoppers to control the rate of crack growth, and to provide adequate ________.

A. ultimate strength B. residual strength

C. test data D. crack simulation

9. Any rework or repair must not increase ________ that degrade component durability.

A. stress concentrations B. corrosion areas

C. structural weight D. cost

10. Design values should be based on ________ to assure a high degree of confidence in these values.

A. sufficient testing B. simple testing

C. vital testing D. engineering experience

11. Which of the following is not true about Material Strength Properties and Material Design Values?

A. Consideration of environmental conditions.

B. Experience based design values.

C. Other material design values.

D. Use of higher design values based on premium selection.

Ⅲ. Fill in the Blanks

1. This AMC sets forth an acceptable means of demonstrating compliance with the provisions related to the validation of the methods used for determination of flight load ________ and ________.
2. Compliance can be shown by analysis supported by ________ test evidence, analysis supported by new test evidence or by test only.
3. Testing must be used to show ________ with the strength and deformation requirements.
4. ________ should be provided for classifications other than New Structure.
5. In the case where ________ is arrested at a readily detectable size following rapid crack growth.
6. The metal material is damaged by the surrounding medium, which is called ________.
7. The test ________ used for material property certification testing should be made from material produced using production processes.
8. ________ maneuvers can prevent the aircraft from stall due to the excessive elevation angle.

Ⅳ. Grammar/Logical Mistakes Correction

1. The normal operating stresses in specific regions of the structure are such a low order that serious damage growth is extremely improbable.
2. All differences between the test article and production article should be account for either by analysis support by test evidence or by testing itself.
3. The nature and extend of tests on complete structures or on portions of the primary structure will depend upon applicable previous design, construction, tests, and service experience, in connect with similar structures.
4. The test article should be full-scale (component or sub-component) and represent that portion of the production aircraft requires test.
5. Detection of damage before it becomes dangerous is the optional control in ensuring the damage-tolerance characteristics of the structure.

Ⅴ. True or False

1. The full name of AMC is Avionics Maintenance Conference.
2. The loading spectrum should be based on measured statistical data of the type derived from government and industry load history studies and on a bold estimate of the anticipated use of the aeroplane.
3. The fail-safe substantiation can be valid as long as it is preferred.
4. Service and test experience of similar in service components that were designed

using different design criteria and methods should demonstrate that the load paths and potential failure modes of the components are equal.

5. It is prescribed that the probability of structural failure is to be maximized.

Ⅵ. Questions to Answer

1. What is the full name of AMC?
2. What are the substantial features of Flight Load Validation?
3. Why do we need Flight Load Validation?
4. What is the relationship between Flight Load Measurement and Flight Load Validation?
5. In which case is the Finite Element Method considered reliable?
6. When shall testing be used to show compliance with the strength and deformation requirements?
7. What can we do to control the rate of crack growth?
8. In those cases where inaccessible and uninspectable blind areas exist, what shall we be concerned about?
9. What is the effect of the Scatter Factor?

Ⅶ. Questions for Discussion

1. Why do regulations often prescribe conservative calculation when discussing the airworthiness?
2. Why do thorough choices have to be made with respect to the load conditions applied?
3. How should justification be provided for any approach/method?
4. How do you understand "loading spectrum"?
5. How can we guarantee that the results we obtained by flight load measurement are acceptable?
6. When the structure supports limit loads, in which case is the deformation acceptable? Why?
7. Can you provide typical examples of material strength properties?
8. Why is it important to use correct material strength properties and material design values?
9. What is the concept of the damage-tolerance (fail-safe) approach?
10. What is damage-tolerance evaluation of structure intended for?
11. What can we get from the damage-tolerance analysis and tests?
12. How to make structure work safer for longer time?

Chapter 4 Airworthiness Standards: Systems

Part I Regulatory Document Text Example

[from Federal Aviation Regulations FAR 25 25. 951—25. 1107]

4. 1 Fuel System

4. 1. 1 General

Each fuel system must be constructed and arranged to ensure a flow of fuel at a rate and pressure established for proper engine and auxiliary power unit functioning under each likely operating condition, including any maneuver for which certification is requested and during which the engine or auxiliary power unit is permitted to be in operation.

Each fuel system must be arranged so that any air which is introduced into the system will not result in (1) power interruption for more than 20 seconds for reciprocating engines; or (2) flameout for turbine engines.

Each fuel system for a turbine engine must be capable of sustained operation throughout its flow and pressure range with fuel initially saturated with water at 80 ℉ and having 0. 75 cc of free water per gallon added and cooled to the most critical condition for icing likely to be encountered in operation.

Each fuel system for a turbine engine powered airplane must meet the applicable fuel venting requirements.

4. 1. 2 Fuel System Lightning Protection

The fuel system must be designed and arranged to prevent the ignition of fuel vapor within the system by (a) direct lightning strikes to areas having a high probability of stroke attachment; (b) swept lightning strokes to areas where swept strokes are highly probable; and (c) corona and streamering at fuel vent outlets.

4. 1. 3 Fuel Flow

Each fuel system must provide at least 100 percent of the fuel flow required under each intended operating condition and maneuver. Compliance must be shown as follows:

(1) Fuel must be delivered to each engine at a pressure within the limits specified in the engine type certificate.

(2) The quantity of fuel in the tank may not exceed the amount established as the unusable fuel supply for that tank plus that necessary to show compliance with this section.

(3) Each main pump must be used that is necessary for each operating condition and attitude for which compliance with this section is shown, and the appropriate emergency pump must be substituted for each main pump so used.

4.1.4 Fuel Tanks: General

Each fuel tank must be able to withstand, without failure, the vibration, inertia, fluid, and structural loads that it may be subjected to in operation.

Flexible fuel tank liners must be approved or must be shown to be suitable for the particular application.

Integral fuel tanks must have facilities for interior inspection and repair.

Fuel tanks within the fuselage contour must be able to resist rupture and to retain fuel, under the inertia forces prescribed for the emergency landing conditions. In addition, these tanks must be in a protected position so that exposure of the tanks to scraping action with the ground is unlikely.

Fuel tank access covers must comply with the following criteria in order to avoid loss of hazardous quantities of fuel:

(1) All covers located in an area where experience or analysis indicates a likely strike must be shown by analysis or tests to minimize penetration and deformation by tire fragments, low energy engine debris, or other likely debris.

(2) All covers must be fire resistant.

For pressurized fuel tanks, a means with fail-safe features must be provided to prevent the buildup of an excessive pressure difference between the inside and the outside of the tank.

4.1.5 Fuel Tank Tests

It must be shown by tests that the fuel tanks, as mounted in the airplane, can withstand, without failure or leakage, the more critical of the pressures resulting from the conditions. In addition, it must be shown by either analysis or tests, that tank surfaces subjected to more critical pressures resulting from the condition, are able to withstand the following pressures: (1) an internal pressure of 3.5 psi; (2) 125 percent of the maximum air pressure developed in the tank from ram effect; (3) fluid pressures developed during maximum limit accelerations, and deflections, of the airplane with a full tank; (4) fluid pressures developed during the most adverse combination of airplane roll and fuel load.

Each metallic tank with large unsupported or unstiffened flat surfaces, whose failure or deformation could cause fuel leakage, must be able to withstand the following test, or its equivalent, without leakage or excessive deformation of the tank walls: (1) each complete tank assembly and its supports must be vibration tested while mounted to simulate the actual installation; (2) the tank assembly must be vibrated for 25 hours at an amplitude of not less than 1/32 of an inch (unless another amplitude is substantiated) while 2/3 filled with water or other suitable test fluid.

Except where satisfactory operating experience with a similar tank in a similar installation is shown, nonmetallic tanks must withstand the test, with fuel at a temperature of 110 ℉. During this test, a representative specimen of the tank must be installed in a supporting structure simulating the installation in the airplane.

For pressurized fuel tanks, it must be shown by analysis or tests that the fuel tanks can withstand the maximum pressure likely to occur on the ground or in flight.

4.1.6 Fuel Tank Installations

Each fuel tank must be supported so that tank loads (resulting from the weight of the fuel in the tanks) are not concentrated on unsupported tank surfaces. In addition:

(1) There must be pads, if necessary, to prevent chafing between the tank and its supports;

(2) Padding must be nonabsorbent or treated to prevent the absorption of fluids;

(3) If a flexible tank liner is used, it must be supported so that it is not required to withstand fluid loads; and

(4) Each interior surface of the tank compartment must be smooth and free of projections that could cause wear of the liner unless: (ⅰ) provisions are made for protection of the liner at these points; or (ⅱ) the construction of the liner itself provides that protection.

Spaces adjacent to tank surfaces must be ventilated to avoid fume accumulation due to minor leakage. If the tank is in a sealed compartment, ventilation may be limited to drain holes large enough to prevent excessive pressure resulting from altitude changes.

The location of each tank must meet the requirements.

No engine nacelle skin immediately behind a major air outlet from the engine compartment may act as the wall of an integral tank.

Each fuel tank must be isolated from personnel compartments by a fumeproof and fuelproof enclosure.

4.1.7 Fuel Tank Ignition Prevention

No ignition source may be present at each point in the fuel tank or fuel tank system where catastrophic failure could occur due to ignition of fuel or vapors. This must be shown by:

(1) Determining the highest temperature allowing a safe margin below the lowest expected autoignition temperature of the fuel in the fuel tanks.

(2) Demonstrating that no temperature at each place inside each fuel tank where fuel ignition is possible will exceed the temperature. This must be verified under all probable operating, failure, and malfunction conditions of each component whose operation, failure, or malfunction could increase the temperature inside the tank.

(3) Demonstrating that an ignition source could not result from each single failure, from each single failure in combination with each latent failure condition not shown to be extremely remote, and from all combinations of failures not shown to be extremely improbable. The effects of manufacturing variability, aging, wear, corrosion, and likely damage must be considered.

The fuel tank installation must include either

(1) Means to minimize the development of flammable vapors in the fuel tanks (in the context of this rule "minimize" means to incorporate practicable design methods to reduce the likelihood of flammable vapors); or

(2) Means to mitigate the effects of an ignition of fuel vapors within fuel tanks such that no damage caused by an ignition will prevent continued safe flight and landing.

4.2 Oil System

4.2.1 General

Each engine must have an independent oil system that can supply it with an appropriate quantity of oil at a temperature not above that safe for continuous operation.

The usable oil capacity may not be less than the product of the endurance of the airplane under critical operating conditions and the approved maximum allowable oil consumption of the engine under the same conditions, plus a suitable margin to ensure system circulation. Instead of a rational analysis of airplane range for the purpose of computing oil requirements for reciprocating engine powered airplanes, the following fuel/oil ratios may be used for airplanes without a reserve oil or oil transfer system, a fuel/oil ratio of 30:1 by volume.

Fuel/oil ratios higher than those prescribed may be used if substantiated by data on actual engine oil consumption.

4.2.2 Oil Tanks

1. Installation

Each oil tank installation must meet the requirements.

Expansion space. Oil tank expansion space must be provided as follows:

(1) Each oil tank used with a reciprocating engine must have an expansion space of not less than the greater of 10 percent of the tank capacity or 0.5 gallon, and each oil tank used with a turbine engine must have an expansion space of not less than 10 percent of the tank capacity.

(2) Each reserve oil tank not directly connected to any engine may have an expansion space of not less than two percent of the tank capacity.

(3) It must be impossible to fill the expansion space inadvertently with the airplane in the normal ground attitude.

2. Filler Connection

Each recessed oil tank filler connection that can retain any appreciable quantity of oil must have a drain that discharges clear of each part of the airplane. In addition, each oil tank filler cap must provide an oil-tight seal.

3. Vent

Oil tanks must be vented as follows:

(1) Each oil tank must be vented from the top part of the expansion space so that venting is effective under any normal flight condition.

(2) Oil tank vents must be arranged so that condensed water vapor that might freeze and obstruct the line cannot accumulate at any point.

4. Outlet

There must be means to prevent entrance into the tank itself, or into the tank outlet, of any object that might obstruct the flow of oil through the system. No oil tank outlet may be enclosed by any screen or guard that would reduce the flow of oil below a safe value at any operating temperature. There must be a shutoff valve at the outlet of each oil tank used with a turbine engine, unless the external portion of the oil system (including the oil tank supports) is fireproof.

5. Flexible Oil Tank Liners

Each flexible oil tank liner must be approved or must be shown to be suitable for the particular application.

4.2.3 Oil Radiators

Each oil radiator must be able to withstand, without failure, any vibration, inertia, and oil pressure load to which it would be subjected in operation.

Each oil radiator air duct must be located so that, in case of fire, flames coming from normal openings of the engine nacelle cannot impinge directly upon the radiator.

4.2.4 Oil Valves

Each oil shutoff must meet the requirements of § 25.1189.

The closing of oil shutoff means may not prevent propeller feathering.

Each oil valve must have positive stops or suitable index provisions in the "on" and "off" positions and must be supported so that no loads resulting from its operation or from accelerated flight conditions are transmitted to the lines attached to the valve.

4.3 Cooling

4.3.1 General

The powerplant and auxiliary power unit cooling provisions must be able to maintain the temperatures of powerplant components, engine fluids, and auxiliary power unit components and fluids within the temperature limits established for these components and fluids, under ground, water, and flight operating conditions, and after normal engine or auxiliary power unit shutdown, or both.

4.3.2 Cooling Test Procedures

Compliance must be shown for the takeoff, climb, en route, and landing stages of flight that correspond to the applicable performance requirements. The cooling tests must be conducted with the airplane in the configuration, and operating under the conditions, that are critical relative to cooling during each stage of flight. For the cooling tests, a temperature is "stabilized" when its rate of change is less than two degrees F per minute.

Temperatures must be stabilized under the conditions from which entry is made into each stage of flight being investigated, unless the entry condition normally is not one during which component and the engine fluid temperatures would stabilize (in which case, operation through the full entry condition must be conducted before entry into the stage of flight being investigated in order to allow temperatures to reach their natural levels at the time of entry). The takeoff cooling test must be preceded by a period during which the powerplant component and engine fluid temperatures are stabilized with the engines at ground idle.

Cooling tests for each stage of flight must be continued until: (1) the component and engine fluid temperatures stabilize; (2) the stage of flight is completed; or (3) an operating limitation is reached.

4.4 Induction System

4.4.1 Air Induction

The air induction system for each engine and auxiliary power unit must supply: (1) the air required by that engine and auxiliary power unit under each operating condition for which certification is requested; and (2) the air for proper fuel metering and mixture distribution with the induction system valves in any position.

Each reciprocating engine must have an alternate air source that prevents the entry of rain, ice, or any other foreign matter.

Air intakes may not open within the cowling, unless: (1) that part of the cowling is isolated from the engine accessory section by means of a fireproof diaphragm; or (2) for reciprocating engines, there are means to prevent the emergence of backfire flames.

For turbine engine powered airplanes and airplanes incorporating auxiliary power units: (1) there must be means to prevent hazardous quantities of fuel leakage or overflow from drains, vents, or other components of flammable fluid systems from entering the engine or auxiliary power unit intake system; and (2) the airplane must be designed to prevent water or slush on the runway, taxiway, or other airport operating surfaces from being directed into the engine or auxiliary power unit air inlet ducts in hazardous quantities, and the air inlet ducts must be located or protected so as to minimize the ingestion of foreign matter during takeoff, landing, and taxiing.

If the engine induction system contains parts or components that could be damaged by foreign objects entering the air inlet, it must be shown by tests or, if appropriate, by analysis that the induction system design can withstand the foreign object ingestion test conditions without failure of parts or components that could create a hazard.

4.4.2 Induction System Icing Protection

1. Reciprocating Engines

Each reciprocating engine air induction system must have means to prevent and eliminate icing. Unless this is done by other means, it must be shown that, in air free of visible moisture at a temperature of 30 ℉, each airplane with altitude engines using: (1) conventional venturi carburetors have a preheater that can provide a heat rise of 120 ℉ with the engine at 60 percent of maximum continuous power; or (2) carburetors tending to reduce the probability of ice formation has a preheater that can provide a heat rise of 100 ℉ with the engine at 60 percent of maximum continuous power.

2. Turbine Engines

Each turbine engine must operate throughout the flight power range of the engine (including idling), without the accumulation of ice on the engine, inlet system components, or airframe components that would adversely affect engine operation or cause a serious loss of power or thrust (ⅰ) under the icing conditions specified in appendix C, and (ⅱ) in falling and blowing snow within the limitations established for the airplane for such operation.

Each turbine engine must idle for 30 minutes on the ground, with the air bleed available for engine icing protection at its critical condition, without adverse effect, in an atmosphere that is at a temperature between 15 ℉ and 30 ℉ (between −9 ℃ and −1 ℃) and has a liquid water content not less than 0.3 grams per cubic meter in the form of drops having a mean effective diameter not less than 20 microns, followed by momentary operation at takeoff power or thrust. During the 30 minutes of idle operation, the engine may be run up periodically to a moderate power or thrust setting in a manner acceptable to the Administrator.

3. Induction System Ducts and Air Duct Systems

Each induction system duct upstream of the first stage of the engine supercharger and of the auxiliary power unit compressor must have a drain to prevent the hazardous accumulation. No drain may discharge where it might cause a fire hazard.

Each induction system duct must be (1) strong enough to prevent induction system failures resulting from normal backfire conditions; and (2) fire-resistant if it is in any fire zone for which a fire-extinguishing system is required, except that ducts for auxiliary power units must be fireproof within the auxiliary power unit fire zone.

Each duct connected to components between which relative motion could exist must have means for flexibility.

For turbine engine and auxiliary power unit bleed air duct systems, no hazard may result if a duct failure occurs at any point between the air duct source and the airplane unit served by the air.

Each auxiliary power unit induction system duct must be fireproof for a sufficient distance upstream of the auxiliary power unit compartment to prevent hot gas reverse flow from burning through auxiliary power unit ducts and entering any other compartment or area of the airplane in which a hazard would be created resulting from the entry of hot gases. The materials used to form the remainder of the induction system duct and plenum chamber of the auxiliary power unit must be capable of resisting the maximum heat conditions likely to occur.

Each auxiliary power unit induction system duct must be constructed of materials that will not absorb or trap hazardous quantities of flammable fluids that could be ignited in the event of a surge or reverse flow condition.

Part II Analysis, Study and Exercises

Words

administrator	[ədˈmɪnɪstreɪtə(r)]	*n*. 管理人员
adverse	[ˈædvɜːs]	*adj*. 不利的；有害的
aging	[ˈeɪdʒɪŋ]	*n*. 老化 *adj*. 变老的；老旧的 *v*. 变老，使苍老
airframe	[ˈeəfreɪm]	*n*. 机身
altitude	[ˈæltɪtjuːd]	*n*. 海拔高度
ambient	[ˈæmbɪənt]	*adj*. 周围的；外界的；环绕的；产生轻松氛围的 *n*. 周围环境；一种背景音乐
amplitude	[ˈæmplɪtjuːd]	*n*. 振幅值
assembly	[əˈsemblɪ]	*n*. 立法机构；会议；议会；装配
atmospheric	[ˌætməsˈferɪk]	*adj*. 大气的，大气层的
attitude	[ˈætɪtuːd]	*n*. 态度，看法
auxiliary	[ɔːgˈzɪlɪərɪ]	*adj*. 辅助的
certificate	[səˈtɪfɪkət]	*n*. 证明书；电影放映许可证；毕业证；结业证；出生（或结婚、死亡）证；凭证；单据；证书 *v*. 发证书给……；用证书证明
chafe	[tʃeɪf]	*v*. 擦痛；擦得红肿；（尤指因受限制而）恼怒，烦恼，焦躁
compartment	[kəmˈpɑːtmənt]	*n*. 隔间；隔层
comply	[kəmˈplaɪ]	*v*. 遵从；服从 ；顺从
component	[kəmˈpəʊnənt]	*n*. 组成部分；成分；组件，元件 *adj*. 组成的；构成的
compressor	[kəmˈpresə]	*n*. 压气机；压缩机

configuration	[kənˌfɪgəˈreɪʃn]	*n*. 配置;结构;外形
contour	[ˈkɒntʊə]	*n*. 外形; 轮廓; 等高线
corrosion	[kəˈrəʊʒn]	*n*. 腐蚀侵蚀
cowling	[ˈkaʊlɪŋ]	*n*. 整流罩
debris	[ˈdebriː]	*n*. 残骸; 碎片; 破片; 残渣; 垃圾; 废弃物
deflection	[dɪˈflekʃn]	*n*. (尤指击中某物后)突然转向,偏斜,偏离
deformation	[ˌdiːfɔːˈmeɪʃn]	*n*. 损形; 变形; 畸形; 破相; 变丑; 残废
deviating	[ˈdiːvɪeɪtɪŋ]	*v*. 偏离,出轨(deviate 的现在分词形式)
drain	[dreɪn]	*v*. 排水;排空;喝光 *n*. 下水道;耗费; 放油嘴
duct	[dʌkt]	*n*. 管子
eliminate	[ɪˈlɪmɪneɪt]	*v*. 消除
enclosure	[ɪnˈkləʊʒə]	*n*. 圈占地; 圈用地; 围场; 圈地; (信中)附件
encounter	[ɪnˈkaʊntə]	*v*. 遇到
endurance	[ɪnˈdjʊərəns]	*n*. 耐受性; 续航时间
evidence	[ˈevɪdəns]	*n*. 证据
exceed	[ɪkˈsiːd]	*v*. 超过;胜过;优于;超出
excessive	[ɪkˈsesɪv]	*adj*. 过分的;过度的
favorable	[ˈfeɪvərəbl]	*adj*. 有利的;良好的;赞成的,赞许的;讨人喜欢的
fireproof	[ˈfaɪəpruːf]	*adj*. 防火的
flameout	[ˈfleɪmaʊt]	*n*. 熄火;惨败;彻底失败;显而易见的失败
fluid	[ˈfluːɪd]	*n*. 流体,液体 *adj*. 流动的;不固定的,易变的;流畅的,优美的;液压传动的
fragment	[ˈfrægmənt]	*n*. 碎片; 片段 *v*. (使)碎裂,破裂,分裂
frequency	[ˈfriːkwənsɪ]	*n*. 频率
fuelproof	[ˈfjuːəlpruːf]	*adj*. 防燃油的
fume	[fjuːm]	*v*. (对……)大为生气,十分恼火; 冒烟; 冒气

		n. 烟雾
fumeproof	[ˈfjuːmpruːf]	adj. 防烟剂的；防油气的
fuselage	[ˈfjuːzəlɑːʒ]	n. 机身
hazard	[ˈhæzəd]	n. 危险;危害
hazardous	[ˈhæzədəs]	adj. 冒险的;危险的;有害的
horizontal	[ˌhɒrɪˈzɒntl]	adj. 水平的
		n. 水平位置；水平线；水平面
idle	[ˈaɪdl]	adj. 闲置的;懒惰的;停顿的
		vi. 无所事事;虚度;空转
		vt. 虚度;使空转
ignite	[ɪgˈnaɪt]	v. (使)燃烧,着火；点燃
ignition	[ɪgˈnɪʃ(ə)n]	n. 点火;着火;发火;发火装置；点燃
impinge	[ɪmˈpɪndʒ]	v. 对……有明显作用(或影响);妨碍;侵犯,侵占;撞击
improbable	[ɪmˈprɒbəbl]	adj. 不大可能真实的;奇异的;荒谬的
incorporating	[ɪnˈkɔːpəreɪtɪŋ]	v. 合并
inertia	[ɪˈnɜːʃə]	n. 缺乏活力；惰性；保守；惯性
ingestion	[ɪnˈdʒestʃən]	n. 摄入,吸入
inlet	[ˈɪnlet]	n. 入口；进气道
install	[ɪnˈstɔːl]	v. 安装；设置；建立(程序)；使就职,任命
integral	[ˈɪntɪgrəl]	adj. 必需的；不可或缺的；作为组成部分的；完整的；完备的
		n. 整体
interfere	[ˌɪntəˈfɪə]	v. 影响
internal	[ɪnˈtɜːnl]	adj. 里面的；体内的；(物体、机构)内部的
isolate	[ˈaɪsəleɪt]	v. (使)隔离,孤立,脱离；使(某物质、细胞等)分离
leakage	[ˈliːkɪdʒ]	n. 泄漏量；漏损量；泄漏；渗漏；泄露
liner	[ˈlaɪnə]	n. 邮轮；衬里；内衬
malfunctioning	[mælˈfʌŋkʃənɪŋ]	adj. 出故障的;不正常工作的；功能不正常
metallic	[məˈtælɪk]	adj. 金属的
micron	[ˈmaɪkrɒn]	n. 微米 $=10^{-6}$ 米
minimize	[ˈmɪnɪmaɪz]	v. 最小化；降至最低
moisture	[ˈmɒɪstʃə]	n. 水分

momentary	[ˈməʊməntrɪ]	*adj*. 短暂的
nacelle	[nəˈsel]	*n*. 气球吊篮;飞机的驾驶员室;飞机的引擎机舱
overfilling	[ˈəʊvəfɪlɪŋ]	*vi*. 满溢;(overfill 的现在分词形式)
		vt. 装太多;把……装得溢出
		n. 过度盛装;过量
overflow	[ˌəʊvəˈfləʊ]	*n*. 溢出
penetration	[ˌpenəˈtreɪʃn]	*n*. 穿透;渗透;进入;打穿
powerplant	[ˈpaʊəplænt]	*n*. 发动机;动力装置;发电站
precede	[prɪˈsiːd]	*v*. 前面是……;以……为先导
preheater	[priːˈhiːtə]	*n*. 预热器
pressurize	[ˈpreʃəraɪz]	*v*. 逼迫;使迫不得已;使(潜艇、飞机等内)保持正常气压
prevent	[prɪˈvent]	*v*. 防止
reciprocate	[rɪˈsɪprəkeɪt]	*v*. 互换;交换;往复运动;回报;报答;酬答
remainder	[rɪˈmeɪndə]	*n*. 其余部分;剩余物
reverse	[rɪˈvɜːs]	*v*. 反向
rupture	[ˈrʌptʃə(r)]	*n*. 破裂;断裂;爆裂;
		v. 使断裂,裂开,破裂;使(友好关系)破裂;使绝交;毁掉(协议)
satisfactorily	[ˌsætɪsˈfæktərəlɪ]	*adv*. 令人满意地
saturate	[ˈsætʃəˌreɪt]	*v*. 浸透;浸润;渗透;使湿透;使大量吸收;使饱和;饱和磁化;饱和充电;充满;使饱享
		n. 饱和脂肪
		adj. 湿透的
slush	[slʌʃ]	*n*. 融雪,雪泥
specified	[ˈspesɪfaɪd]	*v*. 明确规定(specify 的过去分词和过去式)
stabilize	[ˈsteɪbəlaɪz]	*vt*. 使稳固,使安定
		vi. 稳定,安定
substantiated	[səbˈstænʃɪeɪtɪd]	*adj*. 证实的
substituted	[ˈsʌbstɪtjʊtɪd]	*vt*. 代替;替代
		adj. 代替的;取代的
supercharged	[ˈsuːpətʃɑːdʒd]	*adj*. (用增压器)增压的
supercharger	[ˈsjuːpəˌtʃɑːdʒə]	*n*. 增压室;增压器;机械增压系统

surge	[sɜːdʒ]	*v*. 急剧上升
taxiing	[ˈtæksɪɪŋ]	*n*. 滑行
taxiway	[ˈtæksɪweɪ]	*n*. (飞机的)滑行道
transmit	[trænzˈmɪt]	*vt*. 传输;传播;发射;传达;遗传
	trænsˈmɪt]	*vi*. 传输;发射信号
upstream	[ˌʌpˈstriːm]	*adv*. 向(或在)上游;逆流
utilized	[ˈjuːtəlaɪzd]	*v*. 使用;应用
valve	[vælv]	*n*. 阀门
vapor	[ˈveɪpə]	*n*. 蒸汽;汽;雾
		v. 吹牛;吹嘘
vent	[ˈvent]	*v*. 放出;发泄;给……开孔;表达
		n. 通风孔;吐露;出口;发泄
ventilate	[ˈventɪleɪt]	*v*. 使(房间、建筑物等)通风;使通气
ventilation	[ˌventɪˈleɪʃn]	*n*. 通风设备;空气流通
venting	[ˈventɪŋ]	*v*. 表达,发泄,排气,通气
vent	[vent]	*n*. 通风孔;排气口
vibration	[vaɪˈbreɪʃn]	*n*. 震动;颤动;抖动;共鸣;振动
wear	[weə(r)]	*v*. 穿,带;呈现出;留,蓄
		n. 耐久性;穿着;磨损
withstand	[wɪðˈstænd]	*v*. 承受;抵住;顶住;经受住

Phrases & Expressions

adjacent to	邻近;临近;靠近
accessory section	附属部分
air free of visible moisture	无可见水分的空气
air induction system	进气系统
air inlet duct	进气管;进气道
air intake	进气口
alternate air source	备用气源
auxiliary power	辅助电源
backfire flames	回火火焰
color-coding	颜色编码
continuous operation	连续作业;持续运转;连续运转
cooling provisions	冷却规定;冷却设施
cowl flaps	前围襟翼
emergency pump	应急泵
engine nacelle	发动机机舱;发动机短舱

excessive deformation	过度变形
expansion space	扩展空间；膨胀空间
fail-safe	有自动保险装置的；具有自动防止故障性能的
fire hazard	火灾隐患；着火危险
fire resistant	耐火的，耐高温的；抗火物
fire zone	防火区
fire-extinguishing system	灭火系统，消防系统
fireproof diaphragm	防火隔板
flammable fluid	易燃液体；可燃液体
flexible fuel tank liner	挠性油箱衬套；柔性油箱衬套；柔性燃料罐衬里
flight power range of the engine	发动机飞行功率范围
fluid pressure	流体压力；油液压力
foreign object	异物；外来物
foreign object ingestion test	外来物吸入试验；异物吸入试验
fuel feed condition	燃料供给条件；供油条件
fuel flow	燃油流；燃油流量
fuel leakage	漏油
fuel metering	燃油计量
fuel tank vent	油箱通风孔
fuel venting	燃油泄漏；燃油排泄
fuel/oil ratio	燃油/机油比
fuelproof enclosure	防燃油的隔罩
fuselage contour	机身外形
hazardous quantity	有害量；危险量
hazardous accumulation	积聚到危险程度
hazardous condition	危险状态；危险情况
heat exchanger	热交换器
hot weather operation	在热气候条件下的工作
icing protection	防止结冰，结冰防护
ignition prevention	防火；点燃防护
ignition source	火源
in flight	在飞行中
in operation	运行中；生效
index provisions	指数规定
induction system	进气系统

integral fuel tank	整体油箱
interior inspection	内部检查
interior surface	内表面
latent failure	潜在失效
lightning protection	防雷保护;避雷;闪电防护
loss of power or thrust	动力或推力损失
maximum allowable	允许的最大限度
maximum ambient atmospheric temperature	最高环境大气温度;最高外界大气温度
mean effective diameter	平均有效直径
most adverse combination	最不利的组合
most favorable position	最有利姿态;最有利位置
oil consumption	油耗;滑油消耗量
oil radiator	滑油散热器
oil valve	油阀;油活门
oil-tight seal	油封;耐滑油密封件
operating characteristics	运行特性;工作特性
plenum chamber	增压室,通风室
power interruption	供电中断;功率中断
pressurized tanks	加压罐;增压油箱
prevent and eliminate icing	防冰和除冰措施
probable operating conditions	可能的运行状态;可能的运行条件
proper conduct	正确实施
rational analysis	理性分析;理论分析
reciprocating engine	往复式发动机,活塞发动机
sea level conditions	海平面状况;海平面条件
shutoff valve	截止阀;关闭阀;切断阀
simulated flight conditions	模拟的各种飞行条件
subject to	经受;遭受;受到
sustained operation	持续运行;持续工作
system circulation	系统循环
takeoff power	起飞功率
tank compartment	油箱室;油箱舱
taxiing downwind	顺风滑行
turbine engine	涡轮发动机
unusable fuel supply	不可用的燃料供应;不可用油量
usable oil capacity	可用油量;可用油容量;可用滑油量
venturi carburetor	文丘里汽化器

Abbreviations

APU	Auxiliary Power Unit	辅助动力装置
CDL	Configuration Deviation List	构型偏离清单
MEL	Minimum Equipment List	最低设备清单
MMEL	Master Minimum Equipment List	最低主设备清单

Sentence Comprehension

1. Each fuel system for a turbine engine must be capable of sustained operation throughout its flow and pressure range with fuel initially saturated with water at 80 ℉ and having 0.75 cc of free water per gallon added and cooled to the most critical condition for icing likely to be encountered in operation.

   译文：用于涡轮发动机的燃油系统在使用下述状态的燃油时，必须能在其整个流量和压力范围内持续工作：燃油先在 27 ℃ (80 ℉)时用水饱和，并且每 10 升燃油含有 2 毫升添加的游离水(每 1 美加仑含 0.75 毫升)，然后冷却到在运行中很可能遇到的最临界的结冰条件。

2. The likely failure of any heat exchanger using fuel as one of its fluids may not result in a hazardous condition.

   译文：使用燃料作为其流体之一的任何热交换器的可能故障可能不会导致危险状况。

3. If fuel can be pumped from one tank to another in flight, the fuel tank vents and the fuel transfer system must be designed so that no structural damage to the tanks can occur because of overfilling.

   译文：如果飞行中可将燃油从一个油箱泵送到另一个油箱，则油箱通气系统和燃油转输系统的设计必须使油箱结构不会因油过量加注而被损坏。

4. Each fuel tank must be able to withstand, without failure, the vibration, inertia, fluid, and structural loads that it may be subjected to in operation.

   译文：每个燃油箱都必须承受运行中可能遇到的振动、惯性、油液及结构的载荷而不被损坏。

5. Fuel tanks within the fuselage contour must be able to resist rupture and to retain fuel, under the inertia forces prescribed for the emergency landing conditions in § 25.561.

   译文：机身内的燃油箱在受到第 25.561 条所述应急着陆情况的惯性力作用时，必须不易破裂并能保存燃油。

6. Each metallic tank with large unsupported or unstiffened flat surfaces, whose failure or deformation could cause fuel leakage, must be able to withstand the following test, or its equivalent, without leakage or excessive deformation of the

tank walls.

译文： 每个具有大的无支承(或无加强)平面的金属油箱，如果其损坏或变形可能引起漏油，则都必须能承受下列试验或等效试验，而油箱壁不发生漏油或过度变形。

7. Except where satisfactory operating experience with a similar tank in a similar installation is shown, nonmetallic tanks must withstand the test specified in 2nd paragraph (5) of this section, with fuel at a temperature of 110 ℉.

译文： 除非表明安装相似的同类油箱已有满意的使用经验，否则非金属油箱必须经受本部分第二段中第(5)条规定的试验，所用燃油温度为 43 ℃ (110 ℉)。

8. Each fuel tank must be supported so that tank loads (resulting from the weight of the fuel in the tanks) are not concentrated on unsupported tank surfaces.

译文： 每个燃油箱必须有支承，使油箱载荷(由油箱内燃油重量引起)不集中作用在无支承的油箱表面。

9. Each fuel tank must be isolated from personnel compartments by a fumeproof and fuelproof enclosure.

译文： 燃油箱与载人舱的隔离，必须采用防油气及防燃油的隔罩。

10. Demonstrating that an ignition source could not result from each single failure, from each single failure in combination with each latent failure condition not shown to be extremely remote, and from all combinations of failures not shown to be extremely improbable.

译文： 证明点火源不会由每个单点失效、每个单点失效与每个没有表明为概率极小的潜在失效条件的组合，以及没有表明为概率极小的所有失效的组合等引起。

11. Based on the evaluations required by this section, critical design configuration control limitations, inspections, or other procedures must be established, as necessary, to prevent development of ignition sources within the fuel tank system and must be included in the Airworthiness Limitations section of the Instructions for Continued Airworthiness required by § 25.1529.

译文： 根据本节要求的评估，必须根据需要建立关键的设计配置控制限制、检查程序或其他程序，以防止在燃油箱系统内产生点火源，并且必须将其纳入第 25.1529 条所要求的持续适航文件中的适航限制部分。

12. Each oil tank must be vented from the top part of the expansion space so that venting is effective under any normal flight condition.

译文： 每个油箱都必须从扩展空间的顶部进行排气，以使排气在任何正常飞行条件下均有效。

13. Each oil valve must have positive stops or suitable index provisions in the "on" and "off" positions and must be supported so that no loads resulting from its operation or from accelerated flight conditions are transmitted to the lines

attached to the valve.

译文：每个油阀的"开"和"关"位置都必须有正向的止动装置或合适的指示装置，并且必须有支撑，使其在运行或加速飞行条件下所产生的负载不会传递到与阀相连的管路上。

14. If the tests are conducted under conditions deviating from the maximum ambient atmospheric temperature, the recorded powerplant temperatures must be corrected under 3rd and 4th paragraphs of this section.

译文：如果测试在偏离最大环境大气温度的条件下进行，则记录的动力装置温度必须按本部分第(3)和第(4)条进行校正。

15. The takeoff cooling test must be preceded by a period during which the powerplant component and engine fluid temperatures are stabilized with the engines at ground idle.

译文：在起飞的冷却试验之前，发动机必须在地面慢车状态下运转一段时间，使动力装置部件和发动机所用的液体温度达到稳定。

16. The air required by that engine and auxiliary power unit under each operating condition for which certification is requested; and the air for proper fuel metering and mixture distribution with the induction system valves in any position.

译文：在申请合格审定的每种运行条件下，必须能够供给该发动机和辅助动力装置所需的空气量；当进气系统阀处于任一位置时，必须能够供给正常燃油调节和混合比分配所需的空气量。

17. There must be means to prevent hazardous quantities of fuel leakage or overflow from drains, vents, or other components of flammable fluid systems from entering the engine or auxiliary power unit intake system.

译文：必须有措施防止从可燃流体系统的排水管、排气口或其他部件中泄漏或溢流出的有害量流体进入发动机或辅助动力装置的进气系统。

18. Each induction system duct upstream of the first stage of the engine supercharger and of the auxiliary power unit compressor must have a drain to prevent the hazardous accumulation of fuel and moisture in the ground attitude.

译文：处于发动机第一级增压器和辅助动力装置压气机上游的进气系统管道，必须有放液嘴，以防在地面姿态时燃油和水汽积聚到危险程度。

19. The materials used to form the remainder of the induction system duct and plenum chamber of the auxiliary power unit must be capable of resisting the maximum heat conditions likely to occur.

译文：用于制造进气系统管道其他部分和辅助动力装置进气增压室的材料，必须能经受住很可能出现的最热状态。

20. Each main pump must be used that is necessary for each operating condition and attitude for which compliance with this section is showed, and the appropriate emergency pump must be substituted for each main pump so used.

    译文：必须使用符合本节要求的各种运行条件和姿态所需的各种主泵，并且必须用适当的应急泵代替所使用的各种主泵。

21. The air induction system for each engine and auxiliary power unit must supply: (1) the air required by that engine and auxiliary power unit under each operating condition for which certification is requested; and (2) the air for proper fuel metering and mixture distribution with the induction system valves in any position.

    译文：每个发动机和辅助动力装置的进气系统都必须提供：(1)发动机和辅助动力装置在要求认证的每个操作条件下所需的空气；(2)在进气系统阀门处于任何位置时，用于正确燃油计量和混合气分配的空气。

22. Each turbine engine must operate throughout the flight power range of the engine (including idling), without the accumulation of ice on the engine, inlet system components, or airframe components that would adversely affect engine operation or cause a serious loss of power or thrust.

    译文：每台涡轮发动机都必须在发动机的整个飞行功率范围内(包括空转)运行，不得在发动机、进气系统部件或机身部件上积聚会对发动机运行产生不利影响或导致严重功率或推力损失的冰。

Advanced Word Study

approve [ə'pru:v] *v.* 赞成；同意；批准，通过(计划、要求等)；认可；核准

Definitions:

① to have or express a favorable opinion

② to accept as satisfactory hopes

③ to give formal or official sanction to

④ to take a favorable view

Synonyms: accredit, approbate, authorize, certify, clear, confirm, endorse, finalize, formalize, ratify, sanction, warrant

Antonyms: decline, deny, disallow, disapprove, negative, reject, turn down, veto

Usage examples:

① The state has *approved* the building plans, so work on the new school can begin immediately.

② I couldn't *approve* such conduct.

③ She will *approve* the date of the meeting.

④ Congress *approved* the proposed budget.

⑤ The usable oil capacity may not be less than the product of the endurance of the airplane under critical operating conditions and the *approved* maximum allowable oil consumption of the engine under the same conditions, plus a suitable margin to ensure system circulation.

⑥ Each flexible oil tank liner must be *approved* or must be shown to be suitable for the particular application.

eliminate [ɪ'lɪmɪneɪt] *v*. 排除;清除;消除;(比赛中)淘汰;消灭,干掉

Definitions:

① to remove or get rid of something/somebody

② to defeat a person or a team so that they no longer take part in a competition, etc.

③ (formal) to kill somebody, especially an enemy or opponent

Synonyms: extinguish, get rid of, remove

Antonyms: add

Usage examples:

① Credit cards *eliminate* the need to carry a lot of cash.

② The police have *eliminated* two suspects from their investigation.

③ This diet claims to *eliminate* toxins from the body.

④ She was *eliminated* from the tournament in the first round.

⑤ Most of his opponents were *eliminated*.

⑥ Each reciprocating engine air induction system must have means to prevent and *eliminate* icing.

favourable ['feɪvərəbl] *adj*. 给人好印象的;肯定的;赞同的;支持的;有利的;有助于……的

Definitions:

① making people have a good opinion of somebody/something

② positive and showing your good opinion of somebody/something

③ good for something and making it likely to be successful or have an advantage

④ fairly good and not too expensive

Synonyms: advantageous, beneficial, good, nice, upstanding, well, lucky

Antonyms: unfavorable, disadvantageous, bad, unlucky

Usage examples:

① She made a *favourable* impression on his parents.

② Reviews of the book have been *favourable*.

③ The terms of the agreement are *favourable* to both sides.

④ They offered me a loan on very *favourable* terms.

⑤ The airplane must be in the following configuration: (1) landing gear retracted; (2) wing flaps in the most *favourable* position.

⑥ Compliance with the requirements must be shown at each weight, altitude, and ambient temperature within the operational limits established for the airplane and with the most *unfavourable* center of gravity for each configuration.

ignite [ɪgˈnaɪt] *v*. (使)燃烧,着火;点燃

Definitions:

① to set (something) on fire, to cause (something) to burn

② to begin burning, to catch fire

③ to give life or energy to (someone or something)

④ to cause the sudden occurrence of (something)

Synonyms: kindle, fire, burn, light

Antonyms: extinguish, put out

Usage examples:

① The fire was *ignited* by sparks.

② The paper *ignited* on contact with sparks.

③ Three wins in a row *ignited* the team.

④ The story *ignited* her imagination.

⑤ His proposal is *igniting* opposition.

⑥ Each auxiliary power unit induction system duct must be constructed of materials that will not absorb or trap hazardous quantities of flammable fluids that could be *ignited* in the event of a surge or reverse flow condition.

installation [ˌɪnstəˈleɪʃn] *n*. 安装的设备(或机器);安装;设置;设施

Definitions:

① a piece of equipment or machinery that has been fixed in position so that it can be used

② the act of fixing equipment or furniture in position so that it can be used

③ a place where specialist equipment is kept and used

④ the act of placing somebody in a new position of authority, often with a ceremony

Synonyms: installment, facility, initiation, mounting

Antonyms: removal

Usage examples:

① *Installation* of the new system will take several days.

② Each complete tank assembly and its supports must be vibration tested while mounted to simulate the actual *installation*.

③ Except where satisfactory operating experience with a similar tank in a similar *installation* is shown, nonmetallic tanks must withstand the test.

isolate [ˈaɪsəleɪt] *v.* (使)隔离,孤立,脱离;使(某物质、细胞等)分离

Definitions:

① to separate a single substance, cell, etc. from others so that you can study it

② to separate a part of a situation, problem, idea, etc. so that you can see what it is and deal with it separately

③ to separate somebody/something physically or socially from other people or things

Synonyms: insulate, keep apart, set apart

Antonyms: connect, include

Usage examples:

① Researchers are still trying to *isolate* the gene that causes this abnormality.

② It is possible to *isolate* a number of factors that contributed to her downfall.

③ Patients with the disease should be *isolated.*

④ He was immediately *isolated* from the other prisoners.

⑤ Each fuel tank must be *isolated* from personnel compartments by a fumeproof and fuelproof enclosure.

maneuver [məˈnuːvə] *n.* 操控; 特技动作;熟练动作; 机动动作

Definitions:

① a movement performed with care and skill

② a clever plan, action or movement that is used to give somebody an advantage

③ to move or turn skillfully or carefully

④ to control or influence a situation in a skillful but sometimes dishonest way

Synonyms: strategy, trick, control, move

Usage examples:

① You will be asked to perform some standard *maneuvers* during your driving test.

② The amendment was somehow introduced by political *maneuver.*

③ The yachts *maneuvered* for position.

④ There was very little room to *maneuver.*

⑤ Each fuel system must be constructed and arranged to ensure a flow of fuel at a rate and pressure established for proper engine and auxiliary power unit functioning under each likely operating condition, including any *maneuver* for which certification is requested and during which the engine or auxiliary power unit is permitted to be in operation.

⑥ Each fuel system must provide at least 100 percent of the fuel flow required under each intended operating condition and *maneuver.*

provision [prəˈvɪʒn] *n.* 供应，提供；(金钱)预备；供应量，提供物；条文，条款，规定

Definitions:

① the act of supplying somebody with something that they need or want

② something that is supplied

③ preparations that you make for something that might or will happen in the future

④ supplies of food and drink, especially for a long journey

⑤ a condition or an arrangement in a legal document

Synonyms: article, clause, item, term, rule, formulation, order, regulation

Usage examples:

① The government is responsible for the *provision* of health care.

② He had already made *provisions* for his wife and children before the accident.

③ We have enough *provisions* to last us two weeks.

④ Under the *provisions* of the lease, the tenant is responsible for repairs.

⑤ This AMC sets forth an acceptable means of demonstrating compliance with the *provisions* of CS-25 related to material strength properties and material design values.

⑥ The powerplant cooling *provisions* must be able to maintain the temperatures of powerplant components within the temperature limits established for these components under ground, water, and flight operating conditions.

reciprocate [rɪˈsɪprəkeɪt] *v.* 互换；交换；往复运动；回报；报答；酬答

Definitions:

① to give and take mutually

② to return in kind or degree

③ to make a return for something

④ to move forward and backward alternately

Synonyms: recompense, repay, return

Usage examples:

① *Reciprocate* a compliment gracefully.

② We hope to *reciprocate* for your kindness.

③ *Reciprocate* their hospitality by inviting them for a visit.

④ *Reciprocating* valves are used in the system.

⑤ Individuals who have received a dedication are expected to *reciprocate* with a gift, perhaps placing a few folded notes of money into the hat when they give it back.

⑥ For the purpose of computing oil requirements for *reciprocating* engine powered airplanes, the following fuel/oil ratios may be used.

⑦ Each oil tank used with a *reciprocating* engine must have an expansion space.

vent [vent] *n.* 通风孔；吐露；出口；发泄 *v.* 放出；发泄；给……开孔；表达

Definitions:

① an opening through which air, steam, smoke, liquid, etc., can go into or out of a room, machine, or container

② an opportunity or a way to express a strong emotion that you have not openly shown

Synonyms: opening, hole, exhaust

Antonyms: block, surpress

Usage examples:

① She needed to find a *vent* for her frustration.

② A *vent* for the clothes dryer.

③ Each fuel system for a turbine engine powered airplane must meet the applicable fuel *venting* requirements.

④ If fuel can be pumped from one tank to another in flight, the fuel tank *vents* and the fuel transfer system must be designed so that no structural damage to the tanks can occur because of overfilling.

Exercises for Self-Study

Ⅰ. Translate Sentences

1. Fuel must be delivered to each engine at a pressure within the limits specified in the engine type certificate.
2. The quantity of fuel in the tank may not exceed the amount established as the unusable fuel supply for that tank under the requirements of §25.959 plus that necessary to show compliance with this section.
3. The weight of the airplane must be the weight with full fuel tanks, minimum crew, and the ballast necessary to maintain the center of gravity within allowable limits.
4. It must be shown by tests that the fuel tanks, as mounted in the airplane, can withstand, without failure or leakage, the more critical of the pressures resulting from the conditions specified in 1st paragraphs (1) and (2) of this section.
5. Fuel/oil ratios higher than those prescribed in 2nd paragraphs (1) and (2) of this section may be used if substantiated by data on actual engine oil consumption.
6. Each flexible oil tank liner must be approved or must be shown to be suitable for the particular application.
7. Each oil radiator air duct must be located so that, in case of fire, flames coming from normal openings of the engine nacelle cannot impinge directly upon the

radiator.

8. The cooling tests must be conducted with the airplane in the configuration, and operating under the conditions, that are critical relative to cooling during each stage of flight.
9. A fuel tank that is sealed in part of the structural space of an aircraft (etc. wing) to store fuel.

Ⅱ. Multiple Choice Questions

1. They can ________ extremes of temperature and weather without fading or cracking.
   A. overcome B. withstand C. maintain D. preserve
2. The effects of manufacturing variability, ________, ________, ________, and likely damage must be considered.
   A. aging B. wear C. corrosion D. size
3. Each oil tank must be vented from the top part of expansion space so that ________ is effective under normal flight condition.
   A. vent B. venting C. vented D. vents
4. If people who were working are ________, then they have no job or work.
   A. favourable B. idle C. fluid D. ambient
5. Prices range from \$ 119 to \$ 199, depending on the particular ________.
   A. production B. powerplant C. configuration D. nacelle
6. There must be means to prevent hazardous quantities of fuel ________ or overflow from drains, vents, or other components of flammable fluids systems from entering the engine or auxiliary power unit intake system.
   A. emergence B. leakage C. loss D. burning
7. Each reciprocating engine air induction system must have means to prevent and ________ icing.
   A. promote B. cause C. eliminate D. break
8. Each metallic tank with large unsupported or unstiffened flat surfaces, whose failure or deformation could cause ________, must be able to withstand the following test, or its equivalent, without leakage or ________ of the tank walls.
   A. fire hazard B. excessive deformation C. subjection
   D. fuel leakage E. fail-safe
9. Demonstrating that an ________ could not result from each single failure, from each single failure in combination with each latent failure condition, and from all combinations of failures not shown to be ________.
   A. ignition source B. ignition prevention C. ignition
   D. extremely probable E. extremely improbable
10. To some extent, ________ can save fuel.

A. taxiing downwind B. fire hazard

C. descent D. landing configuration

11. The materials used to form the remainder of the induction system duct and ________ of the auxiliary power unit must be capable of resisting the maximum heat conditions likely to occur.

A. fireproof diaphragm B. venture carburetors

C. plenum chamber D. preheaters

12. Each metal can with a large unsupported or unstrengthened flat surface must be able to withstand ________.

A. the specified tests

B. the specified tests or their equivalent requirements

C. the specified tests and their equivalent requirements

Ⅲ. Fill in the Blanks

1. Demonstrating that an ignition source could not result from each single failure and from all combinations of failures not shown to be extremely ________.

2. Each fuel system for a turbine engine powered airplane must meet the applicable fuel venting ________.

3. Fuel must be delivered to each engine at a pressure within the limits specified in the engine ________ certificate.

4. Flexible fuel tank liners must be ________ or must be shown to be suitable for the particular application.

5. For pressurized fuel tanks, a means with fail-safe features must be provided to prevent the buildup of an excessive pressure ________ between the inside and the outside of the tank.

6. During this test, a representative ________ of the tank must be installed in a supporting structure simulating the installation in the airplane.

7. Each interior surface of the tank ________ must be smooth and free of projections that could cause wear of the liner.

8. Each fuel tank must be ________ from personnel compartments by a fumeproof and fuelproof enclosure.

9. Each engine must have an ________ oil system that can supply it with an appropriate quantity of oil.

10. No hazard may result if a duct ________ occurs at any point between the air duct source and the airplane unit served by the air.

Ⅳ. Grammar/Logical Mistakes Correction

1. Each main pump must be used that is necessarily for each operating condition and attitude for witch compliance with this section is shown, and the appropriate emergency pump must be substituted by each main pump.

2. Except where satisfactory operated experience with a similar tank in a similar installation is shown, nonmetallic tanks must withstand the test with fuel at a temperature of 110 ℉.
3. Each oil tank must be venting from the top part of the expansion space so that venting is effective under any normal flight condition.
4. The breather discharge does not constitute a fire hazard if foaming occurs or causes emitted oil striking the pilot's windshield.
5. It must be shown in air free with visible moisture at a temperature of 30 ℉.
6. It must be shown by tests that the fuel tanks can withstand, without failure or leakage, the more critical of the pressures resulting in the conditions specified in paragraphs (1) and (2) of this section.

Ⅴ. True or False

1. Not all the fuel systems for a turbine engine powered airplane must meet the applicable fuel venting requirements.
2. Each induction system duct upstream of the first stage of the engine supercharger and of the auxiliary power unit compressor may or may not have a drain to prevent the hazardous accumulation.
3. The engine nacelle includes engine, air filter, battery, etc.
4. The cooling tests must be conducted with the airplane in the configuration, and operating under the conditions, that are critical relative to cooling during each stage of flight.

Ⅵ. Questions to Answer

1. What shall be done to avoid structural damage of fuel tanks because of overfilling when the fuel is pumped from one tank to another in flight?
2. From which part of the expansion space shall each tank be vented in order for the exhaust to be effective under any normal flight conditions?
3. What shall each induction system duct upstream of the first stage of the engine supercharger and of the auxiliary power unit compressor need to prevent the hazardous accumulation?

Ⅶ. Questions for Discussion

1. Why shall critical design configuration control limitations, inspections, or other procedures be established?
2. What may cause the ignition?
3. Why do you think tanks need expansion space?
4. Why do you think tanks require ventilation systems?
5. Can you think of new methods to eliminate icing during flights?
6. If manufacturers utilize composite materials to construct fuel tanks, what should be paid attention to?

Chapter 5　Certification Procedures: Type Certificates

Part Ⅰ　Regulatory Document Text Example

[from Federal Aviation Regulations FAR 21 (21.15—21.53)]

5.1　Application for Type Certificate

(a) An application for a type certificate is made on a form and in a manner prescribed by the Administrator and is submitted to the appropriate Aircraft Certification Office.

(b) An application for an aircraft type certificate must be accompanied by a three-view drawing of that aircraft and available preliminary basic data.

(c) An application for an aircraft engine type certificate must be accompanied by a description of the engine design features, the engine operating characteristics, and the proposed engine operating limitations.

5.2　Special Conditions

If the Administrator finds that the airworthiness regulations of this subchapter do not contain adequate or appropriate safety standards for an aircraft, aircraft engine, or propeller because of a novel or unusual design feature of the aircraft, aircraft engine or propeller, he prescribes special conditions and amendments thereto for the product. The special conditions are issued and contain such safety standards for the aircraft, aircraft engine or propeller as the Administrator finds necessary to establish a level of safety equivalent to that established in the regulations.

5.3　Issue of Type Certificates

5.3.1　Issue of Type Certificate: Normal, Utility, Acrobatic, Commuter, and Transport Category Aircraft; Manned Free Balloons; Special Classes of Aircraft; Aircraft Engines; Propellers

An applicant is entitled to a type certificate for an aircraft in the normal, utility, acrobatic, commuter, or transport category, or for a manned free balloon, special class of aircraft, or an aircraft engine or propeller, if—

(a) The product qualifies; or

(b) The applicant submits the type design, test reports, and computations necessary to show that the product to be certificated meets the applicable airworthiness, aircraft noise, fuel venting, and exhaust emission requirements of the Federal Aviation Regulations and any special conditions prescribed by the Administrator, and the Administrator finds—

(1) Upon examination of the type design, and after completing all tests and inspections, that the type design and the product meet the applicable noise, fuel venting, and emissions requirements of the Federal Aviation Regulations, and further finds that they meet the applicable airworthiness requirements of the Federal Aviation Regulations or that any airworthiness provisions not complied with are compensated for by factors that provide an equivalent level of safety; and

(2) For an aircraft, that no feature or characteristic makes it unsafe for the category in which certification is requested.

5.3.2 Issuance of Type Certificate: Primary Category Aircraft

(a) The applicant is entitled to a type certificate for an aircraft in the primary category if:

(1) The aircraft:

(ⅰ) Is unpowered; is an airplane powered by a single, naturally aspirated engine with a 61-knot or less V_{so} stall speed; or is a rotorcraft with a 6-pound per square foot main rotor disc loading limitation, under sea level standard day conditions;

(ⅱ) Weighs not more than 2,700 pounds; or, for seaplanes, not more than 3,375 pounds;

(ⅲ) Has a maximum seating capacity of not more than four persons, including the pilot; and

(ⅳ) Has an unpressurized cabin.

(2) The applicant has submitted:

(ⅰ) A statement, in a form and manner acceptable to the Administrator, certifying that: the applicant has completed the engineering analysis necessary to demonstrate compliance with the applicable airworthiness requirements; the applicant has conducted appropriate flight, structural, propulsion, and systems tests necessary to show that the aircraft, its components, and its equipment are reliable and function properly; the type design complies with the airworthiness standards and noise requirements established for the aircraft; and no feature or characteristic makes it unsafe for its intended use;

(ⅱ) The flight manual, including any information required to be furnished by the applicable airworthiness standards;

(ⅲ) Instructions for continued airworthiness; and

(ⅳ) A report that: summarizes how compliance with each provision of the type certification basis was determined; lists the specific documents in which the type certification data information is provided; lists all necessary drawings and documents used to define the type design; and lists all the engineering reports on tests and computations that the applicant must retain and make available to substantiate compliance with the applicable airworthiness standards.

(3) The Administrator finds that:

(ⅰ) The aircraft complies with those applicable airworthiness requirements; and

(ⅱ) The aircraft has no feature or characteristic that makes it unsafe for its intended use.

(b) An applicant may include a special inspection and preventive maintenance program as part of the aircraft's type design or supplemental type design.

(c) For aircraft manufactured outside of the United States in a country with which the United States has a bilateral airworthiness agreement for the acceptance of these aircraft, and from which the aircraft is to be imported into the United States:

(1) The statement required must be made by the civil airworthiness authority of the exporting country; and

(2) The required manuals, placards, listings, instrument markings, and documents must be submitted in English.

5.3.3 Issue of Type Certificate: Import Products

(a) A type certificate may be issued for a product that is manufactured in a foreign country with which the United States has an agreement for the acceptance of these products for export and import and that is to be imported into the United States if:

(1) The country in which the product was manufactured certifies that the product has been examined, tested, and found to meet:

(ⅰ) The applicable aircraft noise, fuel venting and exhaust emissions requirements of this subchapter, or the applicable aircraft noise, fuel venting and exhaust emissions requirements of the country in which the product was manufactured, and any other requirements the Administrator may prescribe to provide noise, fuel venting and exhaust emission levels no greater than those provided by the applicable aircraft noise, fuel venting, and exhaust emission requirements of this subchapter; and

(ⅱ) The applicable airworthiness requirements of this subchapter, or the applicable airworthiness requirements of the country in which the product was manufactured and any other requirements the Administrator may prescribe to provide a level of safety equivalent to that provided by the applicable airworthiness requirements of this subchapter;

(2) The applicant has submitted the technical data, concerning aircraft noise and airworthiness, respecting the product required by the Administrator; and

(3) The manuals, placards, listings, and instrument markings required by the applicable airworthiness (and noise, where applicable) requirements are presented in the English language.

(b) A product type certificated under this section is considered to be type certificated under the noise standards of part 36, and the fuel venting and exhaust emission standards of Part 34, of the Federal Aviation Regulations where compliance therewith is certified, and under the airworthiness standards of that part of the Federal Aviation Regulations with which compliance is certified or to which an equivalent level of safety is certified.

5.4 Type Design

The type design consists of:

(a) The drawings and specifications, and a listing of those drawings and specifications, necessary to define the configuration and the design features of the product shown to comply with the requirements of that part of this subchapter applicable to the product;

(b) Information on dimensions, materials, and processes necessary to define the structural strength of the product;

(c) The Airworthiness Limitations section of the Instructions for Continued Airworthiness as required by Parts 23, 25, 27, 29, 31, 33, and 35 of this chapter or as otherwise required by the Administrator; and as specified in the applicable airworthiness criteria for special classes of aircraft; and

(d) For primary category aircraft, if desired, a special inspection and preventive maintenance program designed to be accomplished by an appropriately rated and trained pilot-owner.

(e) Any other data necessary to allow, by comparison, the determination of the airworthiness, noise characteristics, fuel venting, and exhaust emissions (where applicable) of later products of the same type.

5.5 Inspection and Tests

(a) Each applicant must allow the Administrator to make any inspection and any flight and ground test necessary to determine compliance with the applicable requirements of the Federal Aviation Regulations. However, unless otherwise authorized by the Administrator:

(1) No aircraft, aircraft engine, propeller, or part thereof may be presented to the Administrator for test unless compliance has been shown for that aircraft, aircraft engine, propeller, or part thereof; and

(2) No change may be made to an aircraft, aircraft engine, propeller, or part thereof between the time that compliance is shown for that aircraft, aircraft engine, propeller, or part thereof and the time that it is presented to the Administrator for test.

(b) Each applicant must make all inspections and tests necessary to determine:

(1) Compliance with the applicable airworthiness, aircraft noise, fuel venting, and exhaust emission requirements;

(2) That materials and products conform to the specifications in the type design;

(3) That parts of the products conform to the drawings in the type design; and

(4) That the manufacturing processes, construction and assembly conform to those specified in the type design.

5.6 Flight Test

5.6.1 Make Flight Tests

(a) Each applicant for an aircraft type certificate must make the tests. Before making the tests the applicant must show:

(1) Compliance with the applicable structural requirements of this subchapter;

(2) Completion of necessary ground inspections and tests;

(3) That the aircraft conforms with the type design; and

(4) That the Administrator received a flight test report from the applicant (signed, in the case of aircraft to be certificated, by the applicant's test pilot) containing the results of his tests.

(b) Upon showing compliance with paragraph (a) of this section, the applicant must make all flight tests that the Administrator finds necessary:

(1) To determine compliance with the applicable requirements of this subchapter; and

(2) For aircraft to be certificated under this subchapter, except gliders and except airplanes of 6,000 lbs. or less maximum certificated weight that are to be certificated, to determine whether there is reasonable assurance that the aircraft, its components, and its equipment are reliable and function properly.

(c)Each applicant must, if practicable, make the tests upon the aircraft that was used to show compliance with: (1) paragraph of this section; and (2) for rotorcraft, the rotor drive endurance tests, as applicable.

(d) Each applicant must show for each flight test (except in a glider or a manned free balloon) that adequate provision is made for the flight test crew for emergency egress and the use of parachutes.

(e)Except in gliders and manned free balloons, an applicant must discontinue flight tests under this section until he shows that corrective action has been taken, whenever: (1) the applicant's test pilot is unable or unwilling to make any of the required flight tests; or (2) items of noncompliance with requirements are found that may make additional test data meaningless or that would make further testing unduly hazardous.

(f) The flight tests must include: (1) for aircraft incorporating turbine engines of a type not previously used in a type certificated aircraft, at least 300 hours of operation with a full complement of engines that conform to a type certificate; and (2) for all other aircraft, at least 150 hours of operation.

5.6.2 Flight Test Pilot

Each applicant for a normal, utility, acrobatic, commuter, or transport category aircraft type certificate must provide a person holding an appropriate pilot certificate to make the flight tests required by this part.

5.6.3 Flight Test Instrument Calibration and Correction Report

(a) Each applicant for a normal, utility, acrobatic, commuter, or transport category aircraft type certificate must submit a report to the Administrator showing the computations and tests required in connection with the calibration of instruments used for test purposes and in the correction of test results to standard atmospheric conditions.

(b) Each applicant must allow the Administrator to conduct any flight tests that he finds necessary to check the accuracy of the report submitted.

5.7 Type Certificate

Each type certificate is considered to include the type design, the operating limitations, the certificate data sheet, the applicable regulations of this subchapter with which the Administrator records compliance, and any other conditions or limitations prescribed for the product in this subchapter.

5.8 Instructions for Continued Airworthiness and Manufacturer's Maintenance Manuals Having Airworthiness Limitations Sections

(a) The holder of a type certificate for a rotorcraft for which a Rotorcraft Maintenance Manual containing an "Airworthiness Limitations" section has been issued, and who obtains approval of changes to any replacement time, inspection interval, or related procedure in that section of the manual, shall make those changes available upon request to any operator of the same type of rotorcraft.

(b) The holder of a design approval, including either the type certificate or supplemental type certificate for an aircraft, aircraft engine, or propeller for which application was made after January 28, 1981, shall furnish at least one set of complete Instructions for Continued Airworthiness, or as specified in the applicable airworthiness criteria for special classes of aircraft, as applicable, to the owner of each type of aircraft, aircraft engine, or propeller upon its delivery, or upon issuance of the first standard airworthiness certificate for the affected aircraft, whichever occurs later, and thereafter make those instructions available to any other person required by this chapter to comply with any of the terms of these instructions. In addition, changes to the Instructions for Continued Airworthiness shall be made available to any person required by this chapter to comply with any of those instructions.

5.9 Duration

A type certificate is effective until surrendered, suspended, revoked, or a termination date is otherwise established by the Administrator.

5.10 Statement of Conformity

(a) Each applicant must submit a statement of conformity to the Administrator for each aircraft engine and propeller presented to the Administrator for type certification. This statement of conformity must include a statement that the aircraft engine or propeller conforms to the type design therefor.

(b) Each applicant must submit a statement of conformity to the Administrator for each aircraft or part thereof presented to the Administrator for tests.

Part II Analysis，Study and Exercises

Words

acrobatic	[ˌækrəˈbætɪk]	*adj*. 特技类的
amendment	[əˈmendmənt]	*n*. (法律、文件的)改动，修正案，修改，修订
applicability	[əˌplɪkəˈbɪlətɪ]	*n*. 适用性；适应性
approve	[əˈpruːv]	*v*. 批准，通过(计划、要求等)；赞成；同意；认可；核准
assurance	[əˈʃʊrəns]	*n*. 保证；担保
authorize	[ˈɔːθəraɪz]	*v*. 批准，认可，授权给，被授权
availability	[əˌveɪləˈbɪlətɪ]	*n*. 可用性；有效性；实用性 ；适用性
bilateral	[ˌbaɪˈlætərəl]	*adj*. 双边的
calibration	[ˌkælɪˈbreɪʃn]	*n*. 标定；校准
category	[ˈkætəgɔːrɪ]	*n*. (人或事物的)类别，种类
certifying	[ˈsɜːtɪfaɪɪŋ]	*n*. 证明
commuter	[kəˈmjuːtə]	*n*. 通勤类，班机
compensate	[ˈkɒmpenseɪt]	*v*. 对……加以补偿(补充)
complement	[ˈkɒmplɪment]	*n*. 补充；补足；使完美；使更具吸引力
compliance	[kəmˈplaɪəns]	*n*. 服从；顺从；遵从
computation	[ˌkɒmpjʊˈteɪʃn]	*n*. 计算；计算过程，计算指令
criteria	[kraɪˈtɪərɪə]	*n*. 标准，条件(criterion 的复数)
designation	[ˌdezɪgˈneɪʃn]	*n*. 指定，名称，指示，选派，确定
determination	[dɪˌtɜːmɪˈneɪʃn]	*n*. 决心；果断；坚定；查明；测定；计算
eligibility	[ˌelɪdʒəˈbɪlətɪ]	*n*. 适任，合格；被选举资格
export	[ˈekspɔːt]	*v*. 出口；输出
		n. 出口；输出；出口产品；输出品
glider	[ˈglaɪdə]	*n*. 滑翔机
governing	[ˈgʌvənɪŋ]	*adj*. 统治的；控制的；管理的；治理的
holder	[ˈhəʊldə]	*n*. 持有者；拥有者；支托(或握持)……之物
import	[ˈɪmpɔːt]	*n*. 进口；输入的产品(或劳务)；(产品、劳务的)进口，输入，引进
		v. 进口；输入；引进
incorporating	[ɪnˈkɔːpəreɪtɪŋ]	*v*. 将……包括在内；包含；吸收
		adj. 合并的；联合的
inspection	[ɪnˈspekʃn]	*n*. 视察；检查；查看；审视

interest	[ˈɪntrəst]	*n*. 兴趣;关注;吸引力;趣味;业余爱好;利息;好处;利益;(企业或公司的)股份;权益;股权;利害关系;利益关系;同行;同业;利害与共者;利益团体
		v. 使感兴趣;使关注
interested	[ˈɪntrəstɪd]	*adj*. 感兴趣的;关心的;表现出兴趣的;有利害关系的;当事人的
issue	[ˈɪʃuː]	*n*. 重要议题;争论的问题;(有关某事的)问题,担忧;发行;分发
		v. 宣布;公布;发出;(正式)发给,供给;(尤指通过正式文件)将……诉诸法律;出版;发表
lbs (pounds)	[paʊndz]	*n*. 磅
listing	[ˈlɪstɪŋ]	*n*. (尤指按字母顺序排列的)表册,目录,列表;(表册上的)位置,项目;清单
manual	[ˈmænjʊəl]	*n*. 使用手册;说明书;指南
		adj. 用手的;手工的;体力的;手动的;手控的;用手操作的
nonconventional	[ˌnɒnkənˈvenʃənəl]	*adj*. 非常规的
normal	[ˈnɔːml]	*n*. 常规飞机,正常类
novel	[ˈnɒvl]	*adj*. 新颖的;与众不同的;珍奇的
placards	[ˈplækaːdz]	*n*. 标语牌;广告牌;招贴;海报
portion	[ˈpɔːʃn]	*n*. 部分;(食物的)一份,一客
		v. 把……分成若干份(或部分)
preliminary	[prɪˈlɪmɪnərɪ]	*adj*. 预备性的;初步的;开始的
		n. 初步行动(或活动);预备性措施
prescribe	[prɪˈskraɪb]	*v*. 给……开(药);让……采用(疗法);开(处方);规定;命令;指示
privilege	[ˈprɪvəlɪdʒ]	*n*. 特殊利益;优惠待遇;(有钱有势者的)特权,特殊待遇;荣幸;荣耀;光荣;免责特权
		v. 给予特权;特别优待
provision	[prəˈvɪʒn]	*n*. 提供;供给;给养;供应品;(为将来做的)准备;(法律文件的)规定,条款
		v. 为……提供所需物品(尤指食物)
respecting	[rɪˈspektɪŋ]	*prep*. 关于
revoke	[rɪˈvəʊk]	*v*. 撤销,撤回;废除,取消
specification	[ˌspesɪfɪˈkeɪʃn]	*n*. 规格;规范;明细单;说明书
subchapter	[sʌbˈtʃæptə]	*n*. 分章,子章

supplemental	[ˌsʌpləˈmentl]	*adj*. 补充的，补足的，追加的
surrender	[səˈrendə]	*v*. 投降；(被迫)放弃，交出 *n*. 投降；屈服；屈从；(尤指在战争等过后)放弃，交出
suspend	[səˈspend]	*v*. 悬；挂；吊；暂停；中止；使暂停发挥作用(或使用等)；延缓；推迟；使暂时停职(或停学等)；悬浮
thereof	[ˌðeəˈʌv]	*adv*. 它的；其
thereto	[ˌðeəˈtu]	*adv*. 附之；随之
three-view	[θriː vjuː]	*n*. 三视图
transferability	[ˌtrænsˌfɜːrəˈbɪlətɪ]	*n*. 可转移性，通用性，互换性
transport	[ˈtrænspɔːt]	*n*. 交通运输系统；交通车辆；运输工具；旅行方式；运输；运送；输送；搬运 *v*. (用交通工具)运输，运送，输送；(以自然方式)运输，输送，传播
utility	[juːˈtɪlətɪ]	*n*. 公用事业；实用；效用；有用 *adj*. 多用途的；多效用的；多功能的

Phrases & Expressions

accompanied by	伴随……；附有
acrobatic category aircraft	特技类飞机
adequate provisions	充足的准备；足够的规定
air induction system	进气系统
air inlet duct	进气道
airworthiness criteria	适航标准
airworthiness limitations section	适航性限制部分
applicable regulation	适用条款
applicable requirement	适用要求
aspirated engine	自然吸气引擎
bilateral airworthiness agreement	双边适航协定；双边适航协议
calibration of instruments	仪器的校准
certificate data sheet	审定数据表
check the accuracy of	检查……的准确性
commuter category aircraft	通勤类飞机
comply with	遵从；服从；顺从
confirm to	符合；依从；遵守；遵照
drawings and specifications	图纸及计划规格书

emergency egress	紧急出口
endurance test	耐久性测试；持续性测试
entitled to	有权(享有),有……资格
equivalent level of safety	同级别的安全程度
exhaust emission	废气排放；尾气排放
file a new application	提出一个新申请
file for	提出;申请
fuel venting	燃料排气
function properly	正常运行
in a manner	某种程度上
in accordance with	依照;依据；一致
in the correction of	修正
inspection interval	检测时距
instrument calibration and correction report	仪器校准校正报告
instrument calibration	仪器校准/标准
instrument markings	仪器标记；仪表标记；仪器记录
intended use	预期用途
loading limitation	负载限制
main rotor disc	主旋翼桨盘
maintenance manual	维修手册
maximum seating capacity	最大载客量
noise characteristics	噪音特性
nonconventional aircraft	非常规飞行器
notify in doing something	以……的形式通知
novel design feature	新颖的设计特色
obtain approval of	获得……的批准
operating limitation	操作极限
preliminary basic data	初步基本数据
preventive maintenance program	预防性维护方案；预防性维护计划
preventive maintenance	预防性维修
replacement part	替换零件
sea level standard day condition	海平面标准日间条件
show compliance with	符合……;表明符合性
special inspection	特殊检查
standard atmosphere conditions	标准大气条件
statement of conformity	符合性声明；一致声明
structural strength	结构强度

substantially complete investigation of compliance	充分完整的合规调研
be submitted to	被提交给
supplemental type design	补充型号设计
termination date	终止日期
three-view drawing	三视图
turbine engine	涡轮发动机
type design	型号设计
unpressurized cabin	非加压舱
utility category aircraft	通用类飞机

Abbreviations

ALI	Airworthiness Limitation Instruction	适航性限制项目
CMR	Certification Maintenance Requirement	维修审定要求
FAA	Federal Aviation Administration	美国联邦航空管理局
FAR	Federal Aviation Regulations	联邦航空条例
NTSB	National Transportation Safety Board	国家运输安全委员会(美国)
TCDS	Type Certificate Data Sheet	证书数据表

Sentence Comprehension

1. An application for a type certificate is made on a form and in a manner prescribed by the Administrator and is submitted to the appropriate Aircraft Certification Office.

   译文：以局方规定的表格和方式申请型号合格证，并向有关的航空器审定办公室提出。

2. If the Administrator finds that the airworthiness regulations of this subchapter do not contain adequate or appropriate safety standards for an aircraft, aircraft engine, or propeller because of a novel or unusual design feature of the aircraft, aircraft engine or propeller, he prescribes special conditions and amendments thereto for the product.

   译文：如果局方发现本节适航条例因航空器、航空器发动机或螺旋桨的新颖或不寻常的设计特征而未包含足够或适当的航空器、航空器发动机或螺旋桨安全标准，局方可规定产品的专用条件及修正案。

3. An application for type certification of a transport category aircraft is effective for 5 years and an application for any other type certificate is effective for 3 years, unless an applicant shows at the time of application that his product requires a longer period of time for design, development, and testing, and the Administrator approves a longer period.

译文：运输类航空器的型号合格证申请书有效期为 5 年，而任何其他类航空器的型号合格证申请书的有效期均为 3 年，除非申请人在申请时表明其产品需要较长时间进行设计、开发和测试，并且局方同意批准更长的有效期。

4. The applicant submits the type design，test reports，and computations necessary to show that the product to be certificated meets the applicable airworthiness，aircraft noise，fuel venting，and exhaust emission requirements of the Federal Aviation Regulations and any special conditions prescribed by the Administrator.

   译文：申请人提交必要的型号设计、试验报告和计算结果，以表明所认证的产品符合联邦航空法规中对适航性、航空器噪声、燃油排放和废气排放的要求，以及局方规定的任何特殊条件。

5. The applicant has submitted the technical data，concerning aircraft noise and airworthiness，respecting the product required by the Administrator；and the manuals，placards，listings，and instrument markings required by the applicable airworthiness (and noise，where applicable) requirements are presented in the English language.

   译文：申请人提交了与航空器噪声和适航有关的技术资料，符合局方要求的产品；并且适用的适航性（包括噪声，如适用）要求所需的手册、标牌、清单和仪表标记均以英文呈现。

6. The drawings and specifications，and a listing of those drawings and specifications，necessary to define the configuration and the design features of the product shown to comply with the requirements of that part of this subchapter applicable to the product.

   译文：定义产品的构型和设计特征所必需的图纸和规范，以及这些图纸和规范的清单，以符合本子章该部分对产品的要求。

7. No change may be made to an aircraft，aircraft engine，propeller，or part thereof between the time that compliance with paragraphs (b)(2) through (b)(4) of this section is shown for that aircraft，aircraft engine，propeller，or part thereof and the time that it is presented to the Administrator for test.

   译文：自证明航空器、航空器发动机、螺旋桨或其中的部分符合本部分(b)(2)至(b)(4)款的规定之时起，至提交给适航当局进行测试之时，不得对该航空器、航空器发动机、螺旋桨或其中的部分作出任何更改。

8. Each applicant for a normal，utility，acrobatic，commuter，or transport category aircraft type certificate must submit a report to the Administrator showing the computations and tests required in connection with the calibration of instruments used for test purposes and in the correction of test results to standard atmospheric conditions.

   译文：凡申请正常类、实用类、特技类、通勤类或运输类航空器型号证书的申请人，必须向局方提交一份报告，说明有关用于测试目的的仪器的校准以及将测试结果修正为标准大气条件所需的计算和测试。

9. Each type certificate is considered to include the type design, the operating limitations, the certificate data sheet, the applicable regulations of this subchapter with which the Administrator records compliance, and any other conditions or limitations prescribed for the product in this subchapter.

译文：型号合格证包括型号设计、运行限制、型号合格证数据单，管理员记录符合情况的本分章适用规定，以及本分章中针对产品规定的任何其他条件或限制。

10. A type certificate is effective until surrendered, suspended, revoked, or a termination date is otherwise established by the Administrator.

译文：型号合格证书在中止、暂停、撤销或局方另行确定终止日期之前有效。

11. This statement of conformity must include a statement that the aircraft engine or propeller conforms to the type design therefor.

译文：此符合性声明必须包括航空器发动机或螺旋桨符合其型号设计的声明。

12. The manuals, placards, listings, and instrument markings required by the applicable airworthiness (and noise, where applicable) requirements are presented in the English language.

译文：适用的适航性(包括噪声，如适用)要求所需的手册、标牌、清单和仪表标记均以英文呈现。

13. The type design consists of any other data necessary to allow, by comparison, the determination of the airworthiness, noise characteristics, fuel venting, and exhaust emissions (where applicable) of later products of the same type.

译文：型号设计包含了通过比较在允许后续同型号产品的适航性、噪声特性、燃料排放和废气排放(如适用)的决定时所需的其他必要数据。

14. The Administrator finds necessary for aircraft to be certificated under this subchapter, except gliders and except airplanes of 6,000 lbs, or less maximum certificated weight that are to be certificated under Part 23 of this chapter, to determine whether there is reasonable assurance that the aircraft, its components, and its equipment are reliable and function properly.

译文：适航当局认为有必要根据本子章对飞机进行认证，以确定是否可以合理保证飞机及其部件和设备是可靠的和功能正常的。但滑翔机和重 6 000 磅的飞机，或最大认证重量小于按照本章第 23 部分认证的飞机除外。

15. The flight tests must include, for aircraft incorporating turbine engines of a type not previously used in a type certificated aircraft, at least 300 hours of operation with a full complement of engines that conform to a type certificate.

译文：飞行测试必须包括：对于装有以前未在认证航空器中使用的涡轮发动机的飞机，必须对全套发动机进行至少 300 小时型号合格审定飞行测试。

Advanced Word Study

amendment [əˈmendmənt] *n.* 修正案;修正;合同修改;变更;修订

Definitions:

① a small change or improvement that is made to a law or a document

② the process of changing a law or a document

③ a statement of a change to the law

Synonyms: correction, emendation, improvement, reformation

Usage examples:

① Parliament passed the bill without further *amendment*.

② She made several minor *amendments* to her essay.

③ If the Administrator finds that the airworthiness regulations of this subchapter do not contain adequate or appropriate safety standards for an aircraft, aircraft engine, or propeller because of a novel or unusual design feature of the aircraft, aircraft engine or propeller, he prescribes special conditions and *amendments* thereto for the product.

④ The applicable requirements of this subchapter that are effective on the date of application for that certificate unless compliance with later effective *amendments* is elected or required under this section.

authorize [ˈɔːθəraɪz] *v.* 批准;授权

Definitions: to give official permission for something, or for somebody to do something

Synonyms: accredit, certify, commission, empower, enable, invest, qualify, permit

Antonyms: disqualify, authorize

Usage examples:

① I can *authorize* payments up to $5,000.

② I have *authorized* him to act for me while I am away.

③ Unless otherwise *authorized* by the Administrator.

calibration [ˌkælɪˈbreɪʃn] *n.* 标定;校准;(温度计或其他仪表上的)刻度

Definitions:

① the act of calibrating

② the units of measurement marked on a thermometer or other instrument

Synonyms: standardization, regulation, scale, adjustment, coordination

Usage examples:

① A *calibration* error may affect test results.

② Each applicant for a normal, utility, acrobatic, commuter, or transport category aircraft type certificate must submit a report to the Administrator showing the computations and tests required in connection with the *calibration* of instruments used for test purposes and in the correction of test results to standard atmospheric conditions.

certify [ˈsɜːtɪfaɪ] *v.* (尤指书面)证明,证实;颁发(或授予)专业合格证书;证明(某人)患有精神病

Definitions:

① to state officially, especially in writing, that something is true

② to give somebody an official document proving that they are qualified to work in a particular profession

Synonyms: demonstrate, prove, testify, establish

Usage examples:

① The accounts were *certified* as correct by the finance department.

② The plants must be *certified* to be virus free.

③ He was *certified* as a teacher in 2009.

④ Except as provided by paragraph (c) of this section, a statement, in a form and manner acceptable to the Administrator, *certifying* that: the applicant has completed the engineering analysis necessary to demonstrate compliance with the applicable airworthiness requirements.

designate [ˈdezɪgneɪt] *v.* 命名;指定;选定,指派,委任(某人任某职);标明;标示;指明

Definitions:

① to say officially that somebody/something has a particular character or name

② to describe somebody/something in a particular way

③ to choose or name somebody for a particular job or position

Synonyms: appoint, fix, name, set

Usage examples:

① This area has been *designated* as a National Park.

② This floor has been *designated* a no-smoking area.

③ Several pupils were *designated* as having moderate or severe learning difficulties.

④ *Designated* seats for the elderly are provided.

⑤ The director is allowed to *designate* his/her successor.

⑥ The different types are *designated* by the letters A, B, and C.

⑦ The applicable aircraft noise, fuel venting and exhaust emissions requirements of this subchapter as *designated* in §21.17, and any other requirements the Administrator may prescribe to provide.

entitle [ɪn'taɪtl] *v.* 使享有权利;使符合资格;给……命名(或题名);给(某人)权利(或资格);给……题名;称呼

Definitions:

① to give somebody the right to have or to do something

② to give a title to a book, play, etc.

Synonyms: authorize, privilege, qualify, name

Antonyms: disqualify

Usage examples:

① You will be *entitled* to your pension when you reach 65.

② This ticket does not *entitle* you to travel first class.

③ He read a poem *entitled* "Salt".

④ An applicant is *entitled* to a type certificate for an aircraft in the normal, utility, acrobatic, commuter, or transport category, or for a manned free balloon, special class of aircraft, or an aircraft engine or propeller.

equivalent [ɪ'kwɪvələnt] *adj.* (价值、数量、意义、重要性等)相等的,相同的
n. 相等的东西;等量;对应词

Definitions:

① equal in value, amount, meaning, importance, etc.

② having logical equivalence

③ corresponding or virtually identical especially in effect or function

Synonyms: same, identical, very, equal

Antonyms: different

Usage examples:

① Eight kilometres is roughly *equivalent* to five miles.

② The new regulation was seen as *equivalent* to censorship.

③ Both groups declared *equivalent* statements.

④ A product type certificated under this section is considered to be type certificated under the airworthiness regulations with which compliance is certified or to which an *equivalent* level of safety is certified.

file [faɪl] *n.* 文件夹;卷宗;(计算机的)文件;档案
v. 把(文件等)归档;提起(诉讼);提出(申请);送交(备案);发送(报道给报社)

Definitions:

① a box or folded piece of card for keeping loose papers together and in order

② a collection of information stored together in a computer, under a particular name

③ to arrange in order for preservation and reference

④ to place among official records as prescribed by law

⑤ to initiate (something, such as a legal action) through proper formal procedure

⑥ to register as a candidate especially in a primary election

⑦ to place items in a file

⑧ to submit documents necessary to initiate a legal proceeding

Synonyms: (*n.*) document, paper (*v.*) apply for

Usage examples:

① A stack of *files* awaited me on my desk.

② Every *file* on the same disk must have a different name.

③ *File* a new application for a type certificate and comply with all the provisions of paragraph (a) of this section applicable to an original application.

④ *File* for an extension of the original application and comply with the applicable airworthiness requirements of this subchapter.

nonconventional [ˈnɒnkənˈvenʃənl] *adj.* 非传统的;不依惯例的;非常规的

(opposite to: **conventional** [kənˈvenʃənl] *adj.* 符合习俗的,传统的;常见的;惯例的)

Definitions for **conventional**:

① tending to follow what is done or considered acceptable by society in general

② normal and ordinary, and perhaps not very interesting

③ following what is traditional or the way something has been done for a long time

Synonyms: current, customary, popular, prevailing, prevalent, standard, usual

Antonyms: nonstandard, unconventional, unpopular, unusual, nonconventional

Usage examples:

① *Conventional* behaviour/morality is accepted in the society.

② She's very *conventional* in her views.

③ Researchers used *conventional* methods/approaches.

④ It's not a hotel, in the *conventional* sense, but rather a whole village turned into a hotel.

⑤ You can use a microwave or cook it in a *conventional* oven.

⑥ For special classes of aircraft, including the engines and propellers installed thereon (e. g. , gliders, airships, and other *nonconventional* aircraft), for which airworthiness standards have not been issued under this subchapter, the applicable requirements will be the portions of those other airworthiness requirements contained in Parts 23, 25, 27, 29, 31, 33, and 35 found by the Administrator to be appropriate for the aircraft and applicable to a specific type design.

retain [rɪ'teɪn] *v.* 保持;持有;保留;继续拥有;继续容纳;聘请(律师等)

Definitions:

① to keep something

② to continue to have something

③ to continue to hold or contain something

④ to keep something in one's mind

Synonyms: conserve, withhold, preserve, keep

Antonyms: let go

Usage examples:

① To *retain* your independence.

② He struggled to *retain* control of the situation.

③ A soil that *retains* moisture.

④ I cannot *retain* so much information.

⑤ The president *retained* her as his chief adviser.

⑥ All the engineering reports on tests and computations that the applicant must *retain* and make available to substantiate compliance with the applicable airworthiness standards.

submit [səb'mɪt] *v.* 提交,呈递(文件、建议等);顺从;屈服;投降;不得已接受;表示;认为;主张;建议

Definitions:

① to give a document, proposal, etc. to somebody in authority so that they can study or consider it

② to accept the authority, control or greater strength of somebody/something

Synonyms: yield, give in, present, hand in, suggest

Antonyms: resist

Usage examples:

① to *submit* an application/a claim/a complaint

② Each applicant for aircraft type certificate must *submit* a report to the Administrator.

③ She *submitted* that the local authority should bear the cost.

Exercises for Self-Study

Ⅰ. Translate Sentences

1. An application for an aircraft type certificate must be accompanied by a three-view drawing of that aircraft and available preliminary basic data.
2. If an applicant elects to comply with an amendment to this subchapter that is

effective after the filing of the application for a type certificate, he must also comply with any other amendment that the Administrator finds is directly related.

3. The applicable airworthiness requirements contained in parts 23, 27, 31, 33, and 35 of this subchapter, or such other airworthiness criteria as the Administrator may find appropriate and applicable to the specific design and intended use and provide a level of safety acceptable to the Administrator.
4. An applicant may include a special inspection and preventive maintenance program as part of the aircraft's type design or supplemental type design.
5. Any other data necessary to allow, by comparison, the determination of the airworthiness, noise characteristics, fuel venting, and exhaust emissions (where applicable) of later products of the same type.
6. Each applicant must allow the Administrator to make any inspection and any flight and ground test necessary to determine compliance with the applicable requirements of the Federal Aviation Regulations.
7. ... For aircraft incorporating turbine engines of a type not previously used in a type certificated aircraft, at least 300 hours of operation with a full complement of engines that conform to a type certificate.
8. Each applicant for a normal, utility, acrobatic, commuter, or transport category aircraft type certificate must provide a person holding an appropriate pilot certificate to make the flight tests required by this part.
9. Each applicant must submit a statement of conformity (FAA Form 317) to the Administrator for each aircraft engine and propeller presented to the Administrator for type certification.
10. The type design consists of the drawings and specifications, and a listing of those drawings and specifications necessary to define the configuration and the design features of the product shown to comply with the requirements of that part of this subchapter applicable to the product.

Ⅱ. Multiple Choice Questions

1. The ________ requirements of this subchapter that are effective on the date of application.

   A. substantial　　B. applicable　　C. novel　　D. stupid

2. If valves are periodically inspected and ________ done, the valve will last longer and operate better.

   A. preventive maintenance　　B. special design

   C. relevant reports　　D. replacement

3. Which of the following may be included in the type design?

A. The drawings and specifications, and a listing of those drawings and specifications, necessary to define the configuration and the design features of the product.

B. Information on dimensions, materials, and processes necessary to define the structural strength of the product.

C. A special inspection and preventive maintenance program designed to be accomplished by an appropriately rated and trained pilot-owner.

D. The test instrument calibration and correction report.

4. Any other data necessary to allow the ________ of the airworthiness, noise characteristics, fuel venting, and exhaust emissions of a product.

A. decide B. determination C. decision D. determine

5. For aircraft ________ turbine engines of a type not previously used in a ________ aircraft, at least 300 hours of operation with a full complement of engines that conform to a type certificate.

A. type certificated B. type certification C. fitting D. incorporating

Ⅲ. Fill in the Blanks

1. An applicant may include a special inspection and preventive ________ program as part of the aircraft's type design.

2. A type certificate may be issued for a product that is manufactured in a foreign country with which the United States has an agreement for the ________ of these products for export and import.

3. The applicant has submitted the ________ data concerning aircraft noise and airworthiness.

4. The Airworthiness Limitations section of the ________ for Continued Airworthiness as required by the Administrator.

5. A special inspection and preventive maintenance program designed to be ________ by an appropriately rated and trained pilot-owner.

6. Each applicant must allow the Administrator to conduct any flight tests that he finds necessary to check the ________ of the report submitted.

7. Each type certificate is considered to ________ the type design, the operating limitations, the certificate data sheet, the applicable regulations with which the Administrator records compliance.

Ⅳ. Grammar/Logical Mistakes Correction

1. Application for an aircraft type certificate must accompany with a three-view

drawing of that aircraft and available preliminary basic data.

2. The applicant submits the type design, test reports, and computations necessary shown that the product certificating meets the applicable airworthiness, aircraft noise, fuel venting, and exhaust emission requirements and any special conditions prescribed by the Administrator.
3. A type certificate is effective until surrender, suspend, revoke, or a termination date is otherwise established by the Administrator.
4. Each applicant must allow the Administrator to make any inspection and any flight and ground test necessary to determine compliance of the applicable requirements.
5. An applicant may include a special inspection and preventive maintenance program as part of the aircraft's type design or additional type design.

Ⅴ. True or False

1. Only an application for type certification of acrobatic category aircraft is effective for longer time than an application for any other type certificate, unless an applicant shows that his product requires a longer period of time for design, development, and testing, and the Administrator approves a longer period.
2. An applicant who elects to comply with an amendment to this subchapter that is effective, must also comply with any other amendment that the Administrator finds is directly related.
3. It's necessary for all aircraft to be certificated under transport category to determine whether there is reasonable assurance.
4. The flight tests for aircraft incorporating turbine engines of a type not previously used in a type certificated aircraft must include at least 300 hours of operation with a full complement of engines that conform to a type certificate.
5. A type certificate is effective until surrendered, suspended, revoked, or a termination date is otherwise established by the Administrator.

Ⅵ. Questions to Answer

1. What shall be included as part of aircraft's type design or supplemental type design?
2. Which 3 things are mentioned in the text to show the product meets the applicable airworthiness, aircraft noise, etc. requirements?
3. If the Administrator finds that the airworthiness regulations do not contain adequate or appropriate safety standards for an aircraft because of a novel or unusual design feature, what shall be issued?

Ⅶ. Questions for Discussion

1. What shall reports submitted to the Administrator include?
2. What can the holder of a type certificate for aircraft engines or propellers obtain?
3. Under what circumstances an applicant should apply for a new type certificate?
4. When testing, how many flight hours shall an aircraft need at least?
5. What are the reasons that each applicant shall submit a statement of conformity for?
6. What is the duration of a type certificate?

Chapter 6 Certification Procedures: Airworthiness Certificates

Part I Regulatory Document Text Example

[from Federal Aviation Regulations FAR 21 (21.171—21.199)]

6.1 Applicability

This subpart prescribes procedural requirements for the issue of airworthiness certificates.

6.2 Eligibility

Any registered owner of a U.S.-registered aircraft (or the agent of the owner) may apply for an airworthiness certificate for that aircraft. An application for an airworthiness certificate must be made in a form and manner acceptable to the Administrator, and may be submitted to any FAA office.

6.3 Airworthiness Certificates: Classification

(a) Standard airworthiness certificates are airworthiness certificates issued for aircraft type certificated in the normal, utility, acrobatic, commuter, or transport category, and for manned free balloons, and for aircraft designated by the Administrator as special classes of aircraft.

(b) Special airworthiness certificates are primary, restricted, limited, and provisional airworthiness certificates, special flight permits, and experimental certificates.

6.4 Amendment or Modification

An airworthiness certificate may be amended or modified only upon application to the Administrator.

6.5 Transferability

An airworthiness certificate is transferred with the aircraft.

6.6 Duration

(a) Unless sooner surrendered, suspended, revoked, or a termination date is otherwise established by the Administrator, airworthiness certificates are effective as follows:

(1) Standard airworthiness certificates, special airworthiness certificates—primary category, and airworthiness certificates issued for restricted or limited category aircraft are effective as long as the maintenance, preventive maintenance, and alterations are performed in accordance with Parts 43 and 91 and the aircraft are registered in the United States.

(2) A special flight permit is effective for the period of time specified in the permit.

(3) An experimental certificate for research and development, showing compliance with regulations, crew training, or market surveys is effective for one year after the date of issue or renewal unless a shorter period is prescribed by the Administrator. The duration of amateur-built, exhibition, and air-racing experimental certificates will be unlimited unless the Administrator finds for good cause that a specific period should be established.

(b) The owner, operator, or bailee of the aircraft shall, upon request, make it available for inspection by the Administrator.

(c) Upon suspension, revocation, or termination by order of the Administrator of an airworthiness certificate, the owner, operator, or bailee of an aircraft shall, upon request, surrender the certificate to the Administrator.

6.7 Aircraft Identification

(a) Each applicant for an airworthiness certificate under this subpart must show that his aircraft is identified.

(b) Paragraph (a) of this section does not apply to applicants for the following:

(1) A special flight permit.

(2) An experimental certificate for an aircraft that is not amateur-built or kit-built.

(3) A change from one airworthiness classification to another, for an aircraft already identified.

6.8 Issue of Airworthiness Certificates

6.8.1 Issue of Standard Airworthiness Certificates for Normal, Utility, Acrobatic, Commuter, and Transport Category Aircraft; Manned Free Balloons; and Special Classes of Aircraft

(a) New aircraft manufactured under a production certificate. An applicant for a standard airworthiness certificate for a new aircraft manufactured under a production certificate is entitled to a standard airworthiness certificate without further showing, except that the Administrator may inspect the aircraft to determine conformity to the type design and condition for safe operation.

(b) New aircraft manufactured under type certificate only. An applicant for a standard airworthiness certificate for a new aircraft manufactured under a type certificate only is entitled to a standard airworthiness certificate upon presentation, by the holder or licensee of the type certificate, of the statement of conformity if the Administrator finds after inspection that the aircraft conforms to the type design and is in condition for safe operation.

(c) Import aircraft. An applicant for a standard airworthiness certificate for an import aircraft is entitled to an airworthiness certificate if the country in which the aircraft was manufactured certifies, and the Administrator finds, that the aircraft conforms to the type design and is in condition for safe operation.

(d) Other aircraft. An applicant for a standard airworthiness certificate for aircraft not covered by this section is entitled to a standard airworthiness certificate if:

(1) He presents evidence to the Administrator that the aircraft conforms to a type design approved under a type certificate or a supplemental type certificate and to applicable Airworthiness Directives;

(2) The aircraft (except an experimentally certificated aircraft that previously had been issued a different airworthiness certificate under this section) has been inspected in accordance with the performance rules for 100-hour inspections and found airworthy by: (ⅰ) the manufacturer; (ⅱ) the holder of a repair station certificate as provided in Part 145 of this chapter; (ⅲ) the holder of a mechanic certificate as authorized in Part 65 of this chapter; or (ⅳ) the holder of a certificate issued under Part 121 of this chapter, and having a maintenance and inspection organization appropriate to the aircraft type; and

(3) The Administrator finds after inspection, that the aircraft conforms to the type design, and is in condition for safe operation.

(e) Noise requirements. Notwithstanding all other provisions of this section, the following must be complied with for the original issuance of a standard airworthiness certificate:

(1) For transport category large airplanes and jet (turbojet powered) airplanes that have not had any flight time, no standard airworthiness certificate is originally issued under this section unless the Administrator finds that the type design complies with the noise requirements in addition to the applicable airworthiness requirements in this section. For import airplanes, compliance with this paragraph is shown if the country in which the airplane was manufactured certifies, and the Administrator finds, that the applicable airplane noise requirements of the country in which the airplane was manufactured and any other requirements, the Administrator may prescribe to provide noise levels no greater than those provided, are complied with.

(2) For normal, utility, acrobatic, commuter, or transport category propeller driven small airplanes (except for those airplanes that are designed for "agricultural aircraft operations" or for dispensing fire fighting materials) that have not had any flight time before the applicable date, no standard airworthiness certificate is originally issued under this section unless the applicant shows that the type design complies with the applicable noise requirements of Part 36 in addition to the applicable airworthiness requirements in this section. For import airplanes, compliance with this paragraph is shown if the country in which the airplane was manufactured certifies, and the Administrator finds, that the applicable requirements of Part 36 (or the applicable airplane noise requirements of the country in which the airplane was manufactured and any other requirements the Administrator may prescribe to provide noise levels no greater than those provided by compliance with the applicable requirements of Part 36 of this chapter) are complied with.

(f) Passenger emergency exit requirements. Notwithstanding all other provisions of this section, each applicant for issuance of a standard airworthiness certificate for a transport category airplane manufactured after October 16, 1987, must show that the airplane meets the requirements in effect on July 24, 1989. For the purposes of this paragraph, the date of manufacture of an airplane is the date the inspection acceptance records reflect that the airplane is complete and meets the FAA-approved type design data.

(g) Fuel venting and exhaust emission requirements. Notwithstanding all other provisions of this section, and irrespective of the date of application, no airworthiness certificate is issued, on and after the dates specified in Part 34 for the airplanes specified therein, unless the airplane complies with the applicable requirements of that part.

6.8.2 Issue of Special Airworthiness Certificates for Primary Category Aircraft

(a) New primary category aircraft manufactured under a production certificate. An applicant for an original, special airworthiness certificate-primary category for a new aircraft that meets the criteria, manufactured under a production certificate, including aircraft assembled by another person from a kit provided by the holder of the production certificate and under the supervision and quality control of that holder, is entitled to a special airworthiness certificate without further showing, except that the Administrator may inspect the aircraft to determine conformity to the type design and condition for safe operation.

(b) Imported aircraft. An applicant for a special airworthiness certificate-primary category for an imported aircraft type certificated is entitled to a special airworthiness certificate if the civil airworthiness authority of the country in which the aircraft was manufactured certifies, and the Administrator finds after inspection, that the aircraft conforms to an approved type design that meets the criteria and is in a condition for safe operation.

(c) Aircraft having a current standard airworthiness certificate. An applicant for a special airworthiness certificate-primary category, for an aircraft having a current standard airworthiness certificate that meets the criteria, may obtain the primary category certificate in exchange for its standard airworthiness certificate through the supplemental type certification process. For the purposes of this paragraph, a current standard airworthiness certificate means that the aircraft conforms to its approved normal, utility, or acrobatic type design, complies with all applicable airworthiness directives, has been inspected and found airworthy within the last 12 calendar months, and is found to be in a condition for safe operation by the Administrator.

(d) Other aircraft. An applicant for a special airworthiness certificate-primary category for an aircraft that meets the criteria is entitled to a special airworthiness certificate if:

(1) The applicant presents evidence to the Administrator that the aircraft conforms to an approved primary, normal, utility, or acrobatic type design, including compliance with all applicable airworthiness directives;

(2) The aircraft has been inspected and found airworthy within the past 12 calendar months; and

(3) The aircraft is found by the Administrator to conform to an approved type design and to be in a condition for safe operation.

(e) Multiple-category airworthiness certificates in the primary category and any other category will not be issued; a primary category aircraft may hold only one airworthiness certificate.

6.9 Special Flight Permits

6.9.1 About Special Flight Permits

(a) A special flight permit may be issued for an aircraft that may not currently meet applicable airworthiness requirements but is capable of safe flight, for the following purposes:

(1) Flying the aircraft to a base where repairs, alterations, or maintenance are to be performed, or to a point of storage.

(2) Delivering or exporting the aircraft.

(3) Production flight testing new production aircraft.

(4) Evacuating aircraft from areas of impending danger.

(5) Conducting customer demonstration flights in new production aircraft that have satisfactorily completed production flight tests.

(b) A special flight permit may also be issued to authorize the operation of an aircraft at a weight in excess of its maximum certificated takeoff weight for flight beyond the normal range over water, or over land areas where adequate landing facilities or appropriate fuel is not available. The excess weight that may be authorized under this paragraph is limited to the additional fuel, fuel-carrying facilities, and navigation equipment necessary for the flight.

(c) Upon application, a special flight permit with a continuing authorization may be issued for aircraft that may not meet applicable airworthiness requirements but are capable of safe flight for the purpose of flying aircraft to a base where maintenance or alterations are to be performed. The permit issued under this paragraph is an authorization, including conditions and limitations for flight, which is set forth in the certificate holder's operations specifications. The permit issued under this paragraph may be issued to:

(1) Certificate holders authorized to conduct operations under Part 121 of this chapter; or

(2) Certificate holders authorized to conduct operations under Part 135 for those aircraft they operate and maintain under a continuous airworthiness maintenance program. The permit issued under this paragraph is an authorization, including any conditions and limitations for flight, which is set forth in the certificate holder's operations specifications.

(3) Management specification holders authorized to conduct operations under

Part 91 for those aircraft they operate and maintain under a continuous airworthiness maintenance program.

6.9.2 Issue of Special Flight Permits

(a) An applicant for a special flight permit must submit a statement in a form and manner prescribed by the Administrator, indicating:

(1) The purpose of the flight.

(2) The proposed itinerary.

(3) The crew required to operate the aircraft and its equipment, e. g. , pilot, co-pilot, navigator, etc.

(4) The ways, if any, in which the aircraft does not comply with the applicable airworthiness requirements.

(5) Any restriction the applicant considers necessary for safe operation of the aircraft.

(6) Any other information considered necessary by the Administrator for the purpose of prescribing operating limitations.

(b) The Administrator may make, or require the applicant to make appropriate inspections or tests necessary for safety.

Part II Analysis, Study and Exercises

Words

accordance	[əˈkɔːdns]	*n*. 一致,和谐
acrobatic	[ˌækrəˈbætɪk]	*adj*. 杂技般的; 杂技的;特技的
administrator	[ədˈmɪnɪstreɪtə]	*n*. 管理人员
alteration	[ˌɔːltəˈreɪʃn]	*n*. 交替,变换,改型; 改变; 变化; 更改; 改动
amateur-built	[ˈæmətə(r) bjʊlt]	*adj*. 业余制造的,非专业建造的; 业余建造的,自制的
amend	[əˈmend]	*v*. 修改
amendment	[əˈmendmənt]	*n*. 改动,修正案
applicable	[əˈplɪkəbl]	*adj*. 适用的,合适的
applicant	[ˈæplɪkənt]	*n*. 申请人
assemble	[əˈsemb(ə)l]	*v*. 集合,装配
authorization	[ˌɔːθərəˈzeɪʃn]	*n*. 批准; 授权; 批准书; 授权书
bailee	[beɪˈliː]	*n*. 受寄托人
category	[ˈkætəgərɪ]	*n*. 类别,种类
certify	[ˈsɜːtɪfaɪ]	*v*. 证明,证实

classification	[ˌklæsɪfɪˈkeɪʃn]	*n*. 分类；归类，类别
commuter	[kəˈmjuːtə]	*adj*. 往返的，通勤的
compliance	[kəmˈplaɪəns]	*n*. 服从；顺从
conformity	[kənˈfɔːmətɪ]	*n*. 遵守；符合；一致；遵从
designate	[ˈdezɪgneɪt]	*v*. 标明；标示
directive	[dəˈrektɪv]	*n*. 指令，命令
		adj. 指示的；指导的
dispense	[dɪˈspens]	*v*. 分配；分发；配(药)；发(药)
entitle	[ɪnˈtaɪtl]	*v*. 使享有权利，使符合资格；给(某人)权利(或资格)；给……题名；称呼；使有权利或资格；使有权；给……命名
evacuate	[ɪˈvækjʊeɪt]	*v*. 大逃离；逃亡之路；疏散；撤离；大逃亡
exhaust	[ɪgˈzɔːst]	*n*. 废气
identification	[aɪˌdentɪfɪˈkeɪʃn]	*n*. 确认；验明；身份证明；识别
identify	[aɪˈdentɪfaɪ]	*v*. 识别，确认
impending	[ɪmˈpendɪŋ]	*adj*. 即将发生的；迫在眉睫的
		v. (事件、危险等)逼近；即将临头；吊(在上头)
inspection	[ɪnˈspekʃn]	*n*. 检查，视察
issuance	[ˈɪsjʊəns]	*n*. 发布；核发
itinerary	[aɪˈtɪnərərɪ]	*n*. 行程；旅行日程；航线
kit-built	[kɪt bɪlt]	*n*. 套材组装
limited	[ˈlɪmɪtɪd]	*adj*. 有限的；受(……的)限制的
maintenance	[ˈmeɪntənəns]	*n*. 维护；保养；维持；保持；(依法应负担的)生活费；抚养费
manned	[mænd]	*adj*. 有人控制的，载人的
manufacture	[ˌmænjʊˈfæktʃə]	*n*. 大量制造；批量生产
		v. (用机器)大量生产，成批制造
modification	[ˌmɒdɪfɪˈkeɪʃn]	*n*. 修改；改进
multiple	[ˈmʌltɪpl]	*adj*. 数量多的；多种多样的
normal	[ˈnɔːml]	*n*. 常规飞机
notwithstanding	[nɒtwɪθˈstændɪŋ]	*prep*. 尽管
otherwise	[ˈʌðəwaɪz]	*adv*. 否则；不然；除此以外；在其他方面，亦，或
prescribe	[prɪˈskraɪb]	*v*. 规定
primary	[ˈpraɪmərɪ]	*n*. 初级类
propeller	[prəˈpelə]	*n*. 螺旋桨
provision	[prəˈvɪʒn]	*n*. 提供；供给；给养；供应品；(为将来做的)准备；(法律文件的)规定，条款

		v. 为……提供所需物品(尤指食物)
provisional	[prəˈvɪʒənl]	adj. 临时的；暂时的
register	[ˈredʒɪstə(r)]	v. 登记,注册
renewal	[rɪˈnjuːəl]	n. 有效期延长；更新
request	[rɪˈkwest]	n. (正式或礼貌的)要求,请求
		v. 请求,要求
restrict	[rɪˈstrɪkt]	v. 限制,妨碍
revocation	[ˌrevəˈkeɪʃn]	n. (法律等的)撤销,废除
revoke	[rɪˈvəʊk]	v. 取消；废除；使无效
specify	[ˈspesɪfaɪ]	v. 详细说明;具体说明;明确规定;详述;详列
storage	[ˈstɔːrɪdʒ]	n. 贮存,贮藏(空间)；存储(方式)；付费托管；存放
supervision	[ˌsuːpəˈvɪʒn]	n. 监督,管理
supplemental	[ˌsʌplɪˈmentl]	adj. 补充性的;额外的
surrender	[səˈrendə(r)]	v. (被迫)放弃,交出
suspend	[səˈspend]	v. 中止,暂停
suspension	[səˈspenʃn]	n. 暂停
termination	[ˌtɜːmɪˈneɪʃn]	n. 终止;末端
therein	[ˌðerˈɪn]	adv. 在其中
transfer	[trænsˈfɜː]	v. 转移,(使)调动；转让
transferability	[ˌtrænsˌfɜːrəˈbɪlətɪ]	n. 可转移性;可转让性
turbojet	[ˈtɜːbəʊdʒet]	n. 涡轮喷气飞机
utility	[juːˈtɪlətɪ]	adj. 多效用的；通用的
		n. 多功能飞机；实用;效用;有用
venting	[ˈventɪŋ]	n. 排气,通气
		v. 表达,发泄;给……一个出口;排放;使空气进入;在(上衣)上开衩(vent 的现在分词)

Phrases & Expressions

acrobatic category airplane	特技类飞机
adequate landing facilities	足够的着陆设施
air-racing	空中竞赛
amateur-built	业余建造的,自制的
amended/modified upon	被修改,修正
applicable airworthiness requirement	适航性要求
areas of impending danger	危险区域
authorized to conduct operations	获权操作

commuter category airplane	通勤类飞机
conform to	遵从,服从,与……相符合
crew training	航空人员训练
customer demonstration flights	载客演示飞行;客户示范飞行
determine conformity	确保符合
experimental certificate	实验合格证;实验证书
fire fighting material	消防材料
flight test	试飞
fuel venting and exhaust emission requirements	燃料排气和废气排放要求
in a condition for safe operation	可安全操作状态
in accordance with	按照
irrespective of	不管;不考虑
landing facilities (landing aids)	着陆设施
maintenance and inspection organization	维护和检查组织
management specification	管理规范
manned free balloon	载人气球;载人自由气球
maximum certificated takeoff weight	最大认证起飞重量
mechanic certificate	技师证书
noise level	噪声级别
normal category airplane	正常类飞机
normal range over water	正常水域
passenger emergency exit	乘客紧急出口;客舱紧急出口
point of storage	存放点,贮存处
preventive maintenance	预防性维护
primary category aircraft	初级类航空器
primary category certificate	初级类特殊适航证
propeller-driven aircraft	螺旋桨飞机
set forth	(清晰而有条理地)阐述,陈述,说明
special airworthiness certificate	特殊适航证
special flight permit	特别飞行许可证;特许飞行证
standard airworthiness certificate	标准适航证书
statement of conformity	符合性声明
supplemental type certification process	补充型号合格审定程序
transport category airplane	运输类飞机
type design	型号设计
under the supervision	在……监督下
utility category airplane	实用类飞机

Abbreviations

AD	Airworthiness Directives	适航指令
CAMP	Continuous Airworthiness Maintenance Program	持续适航维护程序
PC	Production Certificate	生产许可证

Sentence Comprehension

1. Standard airworthiness certificates are airworthiness certificates issued for aircraft type certificated in the normal, utility, acrobatic, commuter, or transport category, and for manned free balloons, and for aircraft designated by the Administrator as special classes of aircraft.

   译文: 标准适航证是为获得正常类、实用类、特技类、通勤类或运输类等类别的型号合格证的航空器,载人自由气球以及局方指定的特别类型的航空器颁发的适航证书。

2. Special airworthiness certificates are primary, restricted, limited, and provisional airworthiness certificates, special flight permits, and experimental certificates.

   译文: 特殊适航证分为初级适航证、限制类适航证、限用类适航证、临时适航证、特许飞行证和试验证。

3. The duration of amateur-built, exhibition, and air-racing experimental certificates will be unlimited unless the Administrator finds for good cause that a specific period should be established.

   译文: 由业余爱好者建造的航空器,以及用于展览和空中比赛的航空器的实验证书的期限将是无限的,除非局方有充分理由认为应确定一个特定的有效期。

4. An applicant for a standard airworthiness certificate for a new aircraft manufactured under a production certificate is entitled to a standard airworthiness certificate without further showing, except that the Administrator may inspect the aircraft to determine conformity to the type design and condition for safe operation.

   译文: 根据生产合格证制造的新航空器的标准适航证书的申请人有权获得标准适航证书,但局方人员可以检查航空器,以确定是否符合型号设计和安全操作条件。

5. An applicant for a standard airworthiness certificate for an import aircraft type certificated in accordance with § 21.29 is entitled to an airworthiness certificate if the country in which the aircraft was manufactured certifies, and the Administrator finds, that the aircraft conforms to the type design and is in condition for safe operation.

   译文: 根据第 21.29 条关于进口航空器型号认证的标准,如果飞机制造国证明,并且适航当局认为进口航空器符合型号设计并符合安全运行的条件，适航证书的申请人有权获得标准适航证书。

6. The applicant presents evidence to the Administrator that the aircraft conforms to an approved primary，normal，utility，or acrobatic type design，including compliance with all applicable airworthiness directives.

译文：申请人向局方提供证据，证明航空器符合经批准的初级类、正常类、实用类或特技类型号设计，包括符合所有适用的适航指令。

7. A special flight permit may also be issued to authorize the operation of an aircraft at a weight in excess of its maximum certificated takeoff weight for flight beyond the normal range over water，or over land areas where adequate landing facilities or appropriate fuel is not available.

译文：还可签发特许飞行证，授权在超过航空器的最大认证起飞重量的情况下操作航空器，以便在水面或没有足够着陆设施以及适当燃料的陆地区域飞行。

8. The permit issued under this paragraph is an authorization，including any conditions and limitations for flight，which is set forth in the certificate holder's operations specifications.

译文：根据本条款发出的许可证是一项授权，包括持证人操作规范中规定的飞行条件和限制。

9. Conducting customer demonstration flights in new production aircraft that have satisfactorily completed production flight tests.

译文：在已圆满完成生产飞行试验的新生产的飞机上进行客户示范飞行。

10. Upon application，as prescribed in §§121.79 and 135.17 of this chapter，a special flight permit with a continuing authorization may be issued for aircraft that may not meet applicable airworthiness requirements but are capable of safe flight for the purpose of flying aircraft to a base where maintenance or alterations are to be performed.

译文：根据本章第121.79条和第135.17条的规定，可向可能不符合适航要求但可安全飞行到进行维修或改装的基地的飞机颁发具有持续授权的特别飞行许可证。

11. Management specification holders authorized to conduct operations under Part 91，subpart K，for those aircraft they operate and maintain under a continuous airworthiness maintenance program prescribed by §91.1411 of this part.

译文：根据本部第91.1411条规定的持续适航维护计划，管理规范持有人有权根据第91部分第K分项对他们操作和维护的飞机进行操作。

12. Except as provided in §21.197(c)，an applicant for a special flight permit must submit a statement in a form and manner prescribed by the Administrator，indicating the purpose of the flight.

译文：除第 21.197(c)条规定外，特别飞行许可证申请人须以审定司确定的格式提交一份声明，说明飞行的目的。

13. The ways, if any, in which the aircraft does not comply with the applicable airworthiness requirements.

    译文：飞机不符合适航规定的地方(如有的话)。

14. Unless sooner surrendered, suspended, revoked, or a termination date is otherwise established by the Administrator, airworthiness certificates are effective as follows.

    译文：除适航证项被提前交还、延缓、废除，或失效日期由管理机构另行规定的情况外，以下所列情况下均可认定适航证有效。

15. Notwithstanding all other provisions of this section, and irrespective of the date of application, no airworthiness certificate is issued, on and after the dates specified in Part 34 for the airplanes specified therein, unless the airplane complies with the applicable requirements of that part.

    译文：尽管本部分还有其他规定，但不管申请日期是什么时候，在第 34 部分中所指明的飞机适航证明书当日及之后，均不会核发适航证明书，除非飞机符合该部分适用的规定。

Advanced Word Study

alteration [ˌɔːltəˈreɪʃn] *n*. 改变；变化；更改；改动；交替，变换，改型

Definitions:

① a change in or to something

② the process of changing something

Synonyms: modification, changing

Antonyms: fixation, stabilization

Usage examples:

① Making some simple *alterations* to your diet will make you feel fitter.

② Her jacket was at the boutique waiting for *alteration*.

③ Flying the aircraft to a base where repairs, *alterations*, or maintenance are to be performed, or to a point of storage.

conformity [kənˈfɔːmətɪ] *n*. 遵守；符合；一致；遵从

Definitions: behaviour or actions that follow the accepted rules

Synonyms: conformance, conformation, compliance, accordance

Antonyms: dissimilarity

Usage examples:

① They act in unthinking *conformity* to customs.

② The prime minister is, in *conformity* with their constitution, chosen by the president.

③ An applicant for a standard airworthiness certificate for a new aircraft manufactured under a type certificate only is entitled to a standard airworthiness certificate upon presentation of the statement of *conformity* if the Administrator finds after inspection that the aircraft conforms to the type design and is in condition for safe operation.

duration [djʊˈreɪʃn] *n*. 持续时间;期间

Definitions:

① continuance in time

② the time during which something exists or lasts

③ the length of time that something lasts or continues

Synonyms: life, life span, lifetime, continuance

Usage examples:

① You should gradually increase the *duration* of your workout.

② We were there for the *duration* of the concert.

③ His work contract is of two years' *duration*.

④ The *duration* of amateur-built, exhibition, and air-racing experimental certificates will be unlimited unless the Administrator finds for good cause that a specific period should be established.

issue [ˈɪʃuː] *n*. 重要议题;争论的问题;(有关某事的)问题,担忧;一期;期号
v. 宣布;公布;发出;(正式)发给,供给;(尤指通过正式文件)将……诉诸法律

Definitions:

① an important topic that people are discussing or arguing about

② a problem or worry that somebody has with something

③ one of a regular series of magazines or newspapers

④ the act of supplying or making available things for people to buy or use

⑤ to put forth or distribute usually officially

⑥ to appear or become available through being officially put forth or distributed

Synonyms: (*v.*) put out, print, publish
(*n.*) outcome, distribution, release, issuance

Antonyms: bring about

Usage examples:

① She usually writes about environmental *issues*.

② Money is not an *issue*.

③ The article appeared in *issue* 25.

④ I bought a set of the new stamps on the date of *issue*.

⑤ What is at *issue* is whether she was responsible for her actions.

⑥ *Issue* of special airworthiness certificates for primary category aircraft.

itinerary [aɪˈtɪnərərɪ] *n*. 行程;旅行日程; 航线

Definitions:

① a plan of a trip, including the route and the places that you will visit

② the record of a journey

Synonyms: line, route, thread, travel

Usage examples:

① The next place on our *itinerary* was that city.

② If booking is successful, the customer *itinerary* record is created and its identifier is provided back to the client for future use by the traveler.

③ Except as provided in § 21.197(c), an applicant for a special flight permit must submit a statement in a form and manner prescribed by the Administrator, indicating the proposed *itinerary*.

mechanic [məˈkænɪk] *n*. 机械师;机械修理工;技工;力学;机械学

Definitions:

① of or relating to manual work or skill

② a person whose job is repairing machines or vehicles

③ the science of movement and force

④ the practical study of machinery

⑤ the way something works or is done

Synonyms: mechanical

Antonyms: nonmechanical

Usage examples:

① He has got a *mechanic* certificate.

② The school's car maintenance department where students learn basic *mechanics*.

③ The exact *mechanics* of how payment will be made will be decided later.

④ Flight *mechanics* parameters necessary to correlate the analytical model with flight test results.

⑤ The holder of a *mechanic* certificate as authorized in Part 65 of this chapter.

revocation [ˌrevəˈkeɪʃn] *n*. (法律等的)撤销,废除

Definitions:

① an act or instance of revoking

② the act of cancelling a law, etc.

Synonyms: abandonment, abortion, cancellation, repeal, recall

Antonyms: continuation

Usage examples:

① The *revocation* of planning permission has been issued.

② Upon suspension, *revocation*, or termination by order of the Administrator of an airworthiness certificate, the owner, operator, or bailee of an aircraft shall surrender the certificate to the Administrator.

surrender [sə'rendə] *v.* (被迫)放弃,交出 *n.* 投降,屈从,屈服

Definitions:

① to admit that you have been defeated and want to stop fighting

② to allow yourself to be caught, taken prisoner, etc.

③ to give up something/somebody when you are forced to

④ an act of admitting that you have been defeated and want to stop fighting

⑤ the fact of allowing yourself to be controlled by something

Synonyms: give up

Antonyms: keep, retain, fight back

Usage examples:

① They don't want to *surrender*.

② If we *surrender* the city, the enemy army would surely devastate it all!

③ They had no thought of *surrender*.

④ I believe there is strength in *surrender*.

⑤ Unless sooner *surrendered*, suspended, revoked, or a termination date is otherwise established by the Administrator, airworthiness certificates are effective.

Exercises for Self-Study

Ⅰ. Translate Sentences

1. An airworthiness certificate may be amended or modified only upon application to the Administrator.
2. The owner, operator, or bailee of the aircraft shall, upon request, make it available for inspection by the Administrator.
3. The aircraft (except an experimentally certificated aircraft that previously had been issued a different airworthiness certificate under this section) has been inspected in accordance with the performance rules for 100-hour inspections set forth in § 43.15 of this chapter and found airworthy.
4. The aircraft is found by the Administrator to conform to an approved type design and to be in a condition for safe operation.
5. A special flight permit may be issued for an aircraft that may not currently meet applicable airworthiness requirements but is capable of safe flight.

6. Certificate holders authorized to conduct operations under Part 135 for those aircraft they operate and maintain under a continuous airworthiness maintenance program prescribed by § 135.411 (a)(2) or (b) of that part.
7. Flying the aircraft to a base where repairs, alterations, or maintenance are to be performed, or to a point of storage.
8. The permit issued under this paragraph may be issued to certificate holders authorized to conduct operations under Part 121 of this chapter.
9. The Administrator may make, or require the applicant to make appropriate inspections or tests necessary for safety.
10. Notwithstanding all other provisions of this section, the following must be complied with for the original issuance of a standard airworthiness certificate.

Ⅱ. Multiple Choice Questions

1. An airworthiness certificate is ________ with the aircraft.

   A. changed　B. moved　C. transferred　D. gifted

2. Upon ________, ________, or ________ by order of the Administrator of an airworthiness certificate, the owner or the operator of an aircraft shall surrender the certificate to the Administrator.

   A. suspension　B. revocation　C. termination　D. cancellation

3. Issue of standard airworthiness certificates for ________, ________, ________, commuter, and transport category aircraft.

   A. normal　B. abnormal　C. utility　D. acrobatic

4. Aerobatics refers to the ________ plane along the vertical axis, horizontal axis and vertical axis of space, in a short time to do comprehensive rotation.

   A. normal　B. utility　C. acrobatic　D. stunt

5. The aircraft is ________ to a special airworthiness certificate if it meets the criteria.

   A. assembled　B. entitled　C. allowed　D. permitted

6. The duration of ________, exhibition, and ________ experimental certificates will be unlimited unless the Administrator finds for good cause that a specific period should be established.

   A. amateur-built　B. crew training　C. preventive maintenance
   D. air-racing　E. special classes

7. The aircraft has been inspected in accordance with the performance rules for 100-hour inspections and found airworthy by ________, ________, ________.

   A. the manufacturer
   B. the holder of a repair station certificate
   C. the holder of a mechanic certificate
   D. the Administrator

8. The applicant presents evidence to the Administrator that the aircraft conforms to an approved type design, including compliance with all applicable ________.
   A. special airworthiness certificates B. standard airworthiness certificates
   C. airworthiness directives D. manufacturer specifications
9. Flying the aircraft to a base where repairs, alterations, or maintenance are to be performed, or to ________.
   A. landing facilities B. a point of storage
   C. areas of impending danger D. other countries

Ⅲ. Fill in the Blanks

1. An airworthiness certificate may be ________ or modified only upon application to the Administrator.
2. New aircraft manufactured under ________ certificate only.
3. A special flight permit is ________ for the period of time specified in the permit.
4. The Administrator finds after inspection, that the aircraft ________ the type design, and is in condition for safe operation.
5. The applicant presents ________ to the Administrator that the aircraft conforms to an approved type design.
6. The excess weight that may be ________ under this paragraph is limited to the additional fuel, fuel-carrying facilities, and navigation equipment necessary for the flight.
7. The permit issued under this paragraph is an authorization which is set forth in the ________ holder's operations specifications.

Ⅳ. Grammar/Logical Mistakes Correction

1. Each applicant for issuance of a standard airworthiness certificate for a transport category airplane that is manufactured after October 16, 1987, must show that the airplane satisfies the requirements of § 25.807 that comes into effect on July 24, 1989.
2. An applicant for a special airworthiness certificate, for an aircraft has a current standard airworthiness certificate that meet the criteria may obtain the category in exchange for its standard airworthiness certificate through the supplemental type certification process.
3. The ways, if any, which the aircraft does not comply with the applicable airworthiness requirements.
4. The excess weight that may be authorized includes the additional fuel, fuel-carrying facilities, and navigation equipment necessary for the flight.
5. An applicant for a standard airworthiness certificate for a new aircraft manufactured under a type certificate only is entitled to a standard airworthiness certificate upon presentation of the statement of conformity if the Administrator

finds after inspection that the aircraft conforms for the type design and is in condition for safe operation.

6. An applicant for a standard airworthiness certificate for a new aircraft manufactured under a production certificate is entitled a standard airworthiness certificate without further showing, except that the Administrator may inspect the aircraft to determine conformity to the type design and condition for safe operation.

Ⅴ. True or False

1. Airworthiness certificates only have two kinds: one is standard airworthiness certificate, the other is special airworthiness certificate.
2. An applicant for a special airworthiness certificate for an imported aircraft type certificated is entitled to a special airworthiness certificate (if the CAA of the country in which the aircraft was manufactured certifies, and the Administrator finds after inspection, that the aircraft conforms to an approved type design and is in a condition for safe operation).
3. Aircraft with a temporary nationality certificate can apply for an airworthiness certificate.
4. The duration of amateur-built, exhibition, and air-racing experimental certificates shall be limited.
5. The applicant presents evidence to the Administrator that the aircraft conforms to an approved type design, including compliance with all applicable airworthiness directives.

Ⅵ. Questions to Answer

1. List a few words or phrases that have similar meaning with "notwithstanding" (preposition).
2. What does the abbreviation "dB" mean?
3. In which case shall a special flight permit be issued?
4. What is the full name of FAA?
5. How many airworthiness certificates may a primary category aircraft hold?

Ⅶ. Questions for Discussion

1. What is a standard airworthiness certificate?
2. What is a special airworthiness certificate?
3. What is a production certificate?
4. What is a type certificate?
5. What is the difference between a standard airworthiness certificate and a special airworthiness certificate?
6. Have you ever taken a seat by the emergency exit of an airplane? Do you know what kind of passengers are assigned to emergency exits?
7. What certificates shall a new aircraft obtain?

Chapter 7 Auxiliary Power Unit Installations

Part Ⅰ Regulatory Document Text Example

[from Appendix K to Federal Aviation Regulations FAR Part 25 —Auxiliary Power Unit Installations K25. 901 to K25. 1203]

7. 1 General

7. 1. 1 Installation

(a) For the purpose of this appendix, the Auxiliary Power Unit (APU) includes:

(1) Any engine delivering rotating shaft power, compressed air, or both, which is not intended for direct propulsion of an aircraft;

(2) Each component that affects the control of the APU;

(3) Each component that affects the safety of the APU and the APU installation.

(b) For the purpose of this part:

(1) An essential APU is defined as an APU whose function is required for the dispatch of the airplane and/or continued safe flight;

(2) A non-essential APU is defined as an APU whose function is a matter of convenience, either on the ground or in flight, and may be shut down without jeopardizing safe airplane operation.

(c) For each APU:

(1) The installation must comply with: (ⅰ) the installation instructions provided under the Technical Standard Order (TSO); and (ⅱ) the applicable provisions of this appendix for non-essential APU's; and (ⅲ) the applicable provisions of this appendix for essential APU's.

(2) The components of the installation must be constructed, arranged, and installed so as to ensure their continued safe operation between normal inspections or overhauls;

(3) The installation must be accessible for necessary inspections and maintenance; and

(4) The major components of the installation must be electrically bonded to the other parts of the airplane.

(d) The APU installation must comply with the requirements, except that the effects of the following need not comply with:

(1) APU case burn through or rupture; and

(2) Uncontained APU rotor failure.

7.1.2 Auxiliary Power Unit

(a) Each APU must meet the appropriate requirements of the TSO for its intended function:

(1) Essential: Category Ⅰ APU;

(2) Non-essential: Category Ⅰ or Category Ⅱ APU.

(b) Control of APU rotation and shutdown capability:

(1) It shall be possible to shut down the APU from the flight deck in normal and emergency conditions.

(2) Where continued rotation of an APU could jeopardize the safety of the airplane, there must be a means for stopping rotation. Each component of the stopping system located in the APU compartment must be at least fire resistant.

(c) For APU Installations:

(1) Design precautions must be taken to minimize the hazards to the airplane in the event of an APU rotor failure or of a fire originating within the APU which burns through the APU casing.

(2) The APU system must be designed and installed to give reasonable assurance that those APU operating limitations that adversely affect rotor structural integrity will not be exceeded in service.

(d) Inflight Start Capability:

(1) For non-essential APU's that can be started inflight and all essential APU's: (ⅰ) a means must be provided to start the APU in flight; and (ⅱ) an altitude and airspeed envelope must be established and demonstrated for APU inflight starting.

(2) For Essential APU's, cold soak must be considered in establishing the envelope.

7.1.3 APU Operating Characteristics

(a) APU operating characteristics must be investigated to determine that no adverse characteristics (such as stall, surge, or flame-out) are present, to a hazardous degree, during normal and emergency operation within the range of operation limitations of the airplane and of the APU.

(b) The APU air inlet system may not, as a result of air-flow distortion during normal operation, cause vibration harmful to the APU.

(c) It must be established over the range of operating conditions for which certification is required, that the APU installation vibratory conditions do not exceed the critical frequencies and amplitudes established under the TSO.

7.2 Fuel System

7.2.1 General

(a) Each fuel system must be constructed and arranged to ensure a flow of fuel at a rate and pressure established for proper APU functioning under each likely operating condition, including any maneuver for which certification is requested and during which the APU is permitted to be in operation.

(b) Each fuel system must be arranged so that any air which is introduced into the system will not result in flameout of an essential APU.

(c) Each fuel system for an essential APU must be capable of sustained operation throughout its flow and pressure range with fuel initially saturated with water at 80 ℉ and having 0.75 cc of free water per gallon added and cooled to the most critical condition for icing likely to be encountered in operation.

7.2.2 Fuel Flow

(a) Each fuel system must provide at least 100 percent of the fuel flow required by the APU under each intended operating condition and maneuver. Compliance must be shown as follows:

(1) Fuel must be delivered at a pressure within the limits specified for the APU.

(2) For essential APU's:

(ⅰ) The quantity of fuel in the tank may not exceed the amount established as the unusable fuel supply for that tank plus that necessary to show compliance with this section.

(ⅱ) Each main pump must be used, that is necessary for each operating condition and attitude for which compliance with this section is shown, and the appropriate emergency pump must be substituted for each main pump so used.

(ⅲ) If there is a fuel flowmeter, it must be blocked and the fuel must flow through the meter or its bypass.

(b) If an essential APU can be supplied with fuel from more than one tank, the fuel system must, in addition to having appropriate manual switching capability, be designed to prevent interruption of fuel flow to that APU, without attention by the flight crew, when any tank supplying fuel to that APU is depleted of usable fuel during normal operation, and any other tank, that normally supplies fuel to that APU, contains usable fuel.

7.2.3 Fuel System Hot Weather Operation

For essential APU's:

(a) The fuel supply of an essential APU must perform satisfactorily in hot weather operation. It must be shown that the fuel system from the tank outlet to the APU is pressurized under all intended operations so as to prevent vapor formation. Alternatively, it must be shown that there is no evidence of vapor lock or other malfunctioning during a climb from the altitude of the airport selected by the applicant to the maximum altitude established as an operating limitation, with the APU operating at the most critical conditions for vapor formation but not exceeding the maximum essential load conditions. If the fuel supply is dependent on the same fuel pumps or fuel supply as the main engines, the main engines must be operated at maximum continuous power. The fuel temperature must be at least 110 °F (43 ℃) at the start of the climb.

(b) The test may be performed in flight or on the ground under closely simulated flight conditions. If a flight test is performed in weather cold enough to interfere with the proper conduct of the test, the fuel tank surfaces, fuel lines, and other fuel system parts subject to cold air must be insulated to simulate, insofar as practicable, flight in hot weather.

7.2.4 Fuel System Lines and Fittings

(a) Each fuel line must be installed and supported to prevent excessive vibration and to withstand loads due to fuel pressure and accelerated flight conditions.

(b) Each fuel line connected to components of the airplane between which relative motion could exist must have provisions for flexibility.

(c) Each flexible connection in fuel lines that may be under pressure and subjected to axial loading must use flexible hose assemblies or equivalent means.

(d) Flexible hose must be approved or must be shown to be suitable for the particular application.

(e) No flexible hose that might be adversely affected by exposure to high temperatures may be used where excessive temperatures will exist during operation or after an APU shutdown.

(f) Each fuel line within the fuselage must be designed and installed to allow a reasonable degree of deformation and stretching without leakage.

7.2.5 Fuel Strainer or Filter

For essential APU's, there must be a fuel strainer or filter between the fuel tank outlet and the inlet of either the fuel metering device or an APU-driven positive displacement pump, whichever is nearer the fuel tank outlet. This fuel strainer or filter must:

(a) Be accessible for draining and cleaning and must incorporate a screen or element which is easily removable;

(b) Have a sediment trap and drain except that it need not have a drain if the strainer or filter is easily removable for drain purposes;

(c) Be mounted so that its weight is not supported by the connecting lines or by the inlet or outlet connections of the strainer or filter itself, unless adequate strength margins under all loading conditions are provided in the lines and connections; and

(d) Have the capacity (with respect to operating limitations established for the APU) to ensure that APU fuel system functioning is not impaired, with the fuel contaminated to a degree (with respect to particle size and density) that is greater than that established for the APU under the appropriate TSO.

7.3 Cooling

7.3.1 Cooling Tests

(a) General. Compliance must be shown by tests, under critical conditions. For these tests, the following apply: (1) if the tests are conducted under conditions deviating from the maximum ambient atmospheric temperature, the recorded APU temperatures must be corrected; (2) no corrected temperatures may exceed established limits.

(b) Maximum ambient atmospheric temperature. A maximum ambient atmospheric temperature corresponding to sea level conditions must be established. The temperature lapse rate is 3.6 ℉ (2.0 ℃) per thousand feet of altitude above sea level until a temperature of −69.7 ℉ (−56.5 ℃) is reached, above which altitude, the temperature is considered constant at −69.7 ℉ (−56.5 ℃).

(c) Correction factor. Unless a more rational correction applies, temperatures of APU fluids and components for which temperature limits are established, must be corrected by adding to them the difference between the maximum ambient atmospheric temperature and the temperature of the ambient air at the time of the first occurrence of the maximum component or fluid temperature recorded during the cooling test.

7.3.2 Cooling Test Procedures

(a) Compliance must be shown for the critical conditions that correspond to the applicable performance requirements. The cooling tests must be conducted with the airplane in the configuration, and operating under the conditions that are critical relative to cooling. For the cooling tests, a temperature is "stabilized" when its rate of change is less than 2 ℉ (1 ℃) per minute.

(b) Temperatures must be stabilized prior to entry into each critical condition being investigated, unless the entry condition normally is not one during which component and APU fluid temperatures would stabilize (in which case, operation through the full entry condition must be conducted before entry into the critical condition being investigated in order to allow temperatures to reach their natural levels at the time of entry).

(c) Cooling tests for each critical condition must be continued until:

(1) The component and APU fluid temperatures stabilize;

(2) The stage of flight is completed; or

(3) An operating limitation is reached.

7.4 Air Intake and Bleed Air Duct Systems

7.4.1 Air Intake

The air intake system for the APU:

(a) Must supply the air required by the APU under each operating condition for which certification is requested;

(b) May not draw air from within the APU compartment or other compartments unless the inlet is isolated from the APU accessories and power section by a firewall;

(c) Must have means to prevent hazardous quantities of fuel leakage or overflow from drains, vents, or other components of flammable fluid systems from entering;

(d) Must be designed to prevent water or slush on the runway, taxiway, or other airport operating surface from being directed into the air intake system in hazardous quantities; and

(e) Must be located or protected so as to minimize the ingestion of foreign matter during takeoff, landing, and taxiing.

7.4.2 Air Intake System Ducts

(a) Each air intake system duct must be:

(1) Drained to prevent accumulation of hazardous quantities of flammable fluid and moisture in the ground attitude. The drain(s) must not discharge in locations that might cause a fire hazard; and

(2) Constructed of materials that will not absorb or trap sufficient quantities of flammable fluids such as to create a fire hazard.

(b) Each duct must be:

(1) Designed to prevent air intake system failures resulting from reverse flow, APU surging, or inlet door closure; and

(2) Fireproof within the APU compartment and for a sufficient distance upstream of the APU compartment to prevent hot gas reverse flow from burning through the APU air intake system ducts and entering any other compartment or area of the airplane in which a hazard would be created resulting from the entry of hot gases.

The materials used to form the remainder of the air intake system duct and plenum chamber of the APU must be capable of resisting the maximum heat conditions likely to occur.

(c) Each duct connected to components between which relative motion could exist must have a means for flexibility.

7.4.3 Bleed Air Duct Systems

(a) For APU bleed air duct systems, no hazard may result if a duct failure occurs at any point between the air duct source and the airplane unit served by the bleed air.

(b) Each duct connected to components between which relative motion could exist must have a means for flexibility.

(c) Where the airflow delivery from the APU and main engine is delivered to a common manifold system, precautions must be taken to minimize the possibility of a hazardous condition due to reverse airflow through the APU resulting from malfunctions of any component in the system.

7.5 Exhaust System

7.5.1 General

(a) Each exhaust system must ensure safe disposal of exhaust gases without fire hazard or carbon monoxide contamination in any personnel compartment. For test purposes, any acceptable carbon monoxide detection method may be used to show the absence of carbon monoxide.

(b) Each exhaust system part with a surface hot enough to ignite flammable fluids or vapors must be located or shielded so that leakage from any system carrying flammable fluids or vapors will not result in a fire caused by impingement of the fluids or vapors on any part of the exhaust system including shields for the exhaust system.

(c) Each component that hot exhaust gases could strike, or that could be subjected to high temperatures from exhaust system parts, must be fireproof. All exhaust system components must be separated by fireproof shields from adjacent parts of the airplane that are outside the APU compartment.

(d) No exhaust gases may discharge so as to cause a fire hazard with respect to any flammable fluid vent or drain.

(e) Each exhaust system component must be ventilated to prevent points of excessively high temperature.

(f) Each exhaust shroud must be ventilated or insulated to avoid, during normal operation, a temperature high enough to ignite any flammable fluids or vapors external to the shroud.

7.5.2 Exhaust Piping

(a) Exhaust piping must be heat and corrosion resistant, and must have provisions to prevent failure due to expansion by operating temperatures.

(b) Piping must be supported to withstand any vibration and inertia loads to which it would be subjected in operation.

(c) Piping connected to components between which relative motion could exist must have means for flexibility.

7.6 APU Controls and Accessories

7.6.1 APU Controls

(a) Means must be provided on the flight deck for starting, stopping, and emergency shutdown of each installed APU. Each control must:

(1) Be located, arranged, designed, and marked; and

(2) Be designed and located so that it cannot be inadvertently operated by persons entering, leaving, or moving normally in the flight deck; and

(3) Be able to maintain any set position without constant attention by flight crewmembers and without creep due to control loads or vibration; and

(4) Have sufficient strength and rigidity to withstand operating loads without failure and without excessive deflection; and

(5) For flexible controls, be approved or must be shown to be suitable for the particular application.

(b) APU valve controls located in the flight deck must have:

(1) For manual valves, positive stops or, in the case of fuel valves, suitable index provisions in the open and closed positions;

(2) In the case of valves controlled from the flight deck other than by mechanical means, where the correct functioning of the valve is essential for the safe operation of the airplane, a valve position indicator which senses directly that the valve has attained the position selected must be provided, unless other indications in the flight deck give the flight crew a clear indication that the valve has moved to the selected position. A continuous indicator need not be provided.

(c) For unattended operation, the APU must:

(1) Provide means to automatically shutdown the APU for the following conditions: (ⅰ) exceedance of any APU parameter limit or existence of a detectable hazardous APU operating condition; and (ⅱ) bleed air duct failure between the APU and the airplane unit served by the bleed air, unless it can be shown that no hazard exists to the airplane.

(2) Provide means to automatically shutoff flammable fluids in case of APU compartment fire.

(d) APU controls located elsewhere on the airplane, which are in addition to the flight deck controls, must meet the following requirements:

(1) Each control must be located so that it cannot be inadvertently operated by persons entering, leaving, or moving normally in the area of the control; and

(2) Each control must be able to maintain any set position without creep due to control loads, vibration, or other external forces resulting from the location.

(e) The portion of each APU control located in a designated fire zone that is required to be operated in the event of a fire must be at least fire resistant.

7.6.2 APU Accessories

(a) APU-mounted accessories must be approved for installation on the APU concerned and use the provisions of the APU for mounting.

(b) Electrical equipment subject to arcing or sparking must be installed to minimize the probability of contact with any flammable fluids or vapors that might be present in a free state.

(c) For essential APU's, if continued rotation of a failed aircraft accessory driven by the APU affects the safe operation of the aircraft, there must be means to prevent rotation without interfering with the continued operation of the APU.

7.7 APU Fire Protection

7.7.1 Lines, Fittings and Components

(a) Each line, fitting, and other component carrying flammable fluid in any area subject to APU fire conditions, and each component which conveys or contains flammable fluid in a designated fire zone must be fire resistant, except that flammable fluid tanks and supports in a designated fire zone must be fireproof or be enclosed by a fireproof shield unless damage by fire to any non-fireproof part will not cause leakage or spillage of flammable fluid. Components must be shielded or located to safeguard against the ignition of leaking flammable fluid.

(b) Paragraph (a) of this section does not apply to:

(1) Lines and fittings already approved as part of an APU; and

(2) Vent and drain lines, and their fittings, whose failure will not result in, or add to, a fire hazard.

(c) All components, including ducts, within a designated fire zone must be fireproof if, when exposed to or damaged by fire, they could:

(1) Result in fire spreading to other regions of the airplane; or

(2) Cause unintentional operation of, or inability to operate, essential services or equipment.

7.7.2 Drainage and Ventilation of Fire Zones

(a) There must be complete drainage of each part of each designated fire zone to minimize the hazards resulting from failure or malfunctioning of any component containing flammable fluids. The drainage means must be:

(1) Effective under conditions expected to prevail when drainage is needed; and

(2) Arranged so that no discharged fluid will cause an additional fire hazard.

(b) Each designated fire zone must be ventilated to prevent the accumulation of flammable vapors.

(c) No ventilation opening may be where it would allow the entry of flammable fluids, vapors, or flame from other zones.

(d) Each ventilation means must be arranged so that no discharged vapors will cause an additional fire hazard.

(e) Unless the extinguishing agent capacity and rate of discharge are based on maximum air flow through a zone, there must be means to allow the crew to shutoff sources of forced ventilation to any fire zone.

7.7.3 Shutoff Means

(a) Each APU compartment must have a means to shutoff or otherwise prevent hazardous quantities of flammable fluids, from flowing into, within, or through any designated fire zone, except that shutoff means are not required for:

(1) Lines, fittings and components forming an integral part of an APU; and

(2) Oil systems for APU installations in which all external components of the

oil system, including the oil tanks, are fireproof.

(b) The closing of any fuel shutoff valve for any APU may not make fuel unavailable to the main engines.

(c) Operation of any shutoff may not interfere with the later emergency operation of other equipment.

(d) Each flammable fluid shutoff means and control must be fireproof or must be located and protected so that any fire in a fire zone will not affect its operation.

(e) No hazardous quantity of flammable fluid may drain into any designated fire zone after shutoff.

(f) There must be means to guard against inadvertent operation of the shutoff means and to make it possible for the crew to reopen the shutoff means in flight after it has been closed.

(g) Each tank to APU shutoff valve must be located so that the operation of the valve will not be affected by the APU mount structural failure.

(h) Each shutoff valve must have a means to relieve excessive pressure accumulation unless a means for pressure relief is otherwise provided in the system.

7.7.4 APU Compartment

(a) Each compartment must be constructed and supported so that it can resist any vibration, inertia, and air load to which it may be subjected in operation.

(b) Each compartment must meet the drainage and ventilation requirements.

(c) Each part of the compartment subject to high temperatures due to its nearness to exhaust system parts or exhaust gas impingement must be fireproof.

(d) Each airplane must:

(1) Be designed and constructed so that no fire originating in any APU fire zone can enter, either through openings or by burning through external skin, any other zone or region where it would create additional hazards;

(2) Meet the requirements with the landing gear retracted (if applicable); and

(3) Have fireproof skin in areas subject to flame if a fire starts in the APU compartment.

7.7.5 Extinguishing Agent Containers

(a) Each extinguishing agent container must have a pressure relief to prevent bursting of the container by excessive internal pressures.

(b) The discharge end of each discharge line from a pressure relief connection must be located so that discharge of the fire extinguishing agent would not damage the airplane. The line must be located or protected to prevent clogging caused by ice or other foreign matter.

(c) There must be a means for each fire extinguishing agent container to indicate that the container has discharged or that the charging pressure is below the established minimum necessary for proper functioning.

(d) The temperature of each container must be maintained, under intended operating conditions, to prevent the pressure in the container from:

(1) Falling below that necessary to provide an adequate rate of discharge; or

(2) Rising high enough to cause premature discharge.

(e) If a pyrotechnic capsule is used to discharge the extinguishing agent, each container must be installed so that temperature conditions will not cause hazardous deterioration of the pyrotechnic capsule.

7.7.6 Fire-Detector System

(a) There must be approved, quick acting fire or overheat detectors in each APU compartment in numbers and locations ensuring prompt detection of fire.

(b) Each fire detector system must be constructed and installed so that:

(1) It will withstand the vibration, inertia, and other loads to which it may be subjected in operation;

(2) There is a means to warn the crew in the event that the sensor or associated wiring within a designated fire zone is severed at one point, unless the system continues to function as a satisfactory detection system after the severing; and

(3) There is a means to warn the crew in the event of a short circuit in the sensor or associated wiring within a designated fire zone, unless the system continues to function as a satisfactory detection system after the short circuit.

(c) No fire or overheat detector may be affected by any oil, water, other fluids, or fumes that might be present.

(d) There must be means to allow the crew to check, in flight, the functioning of each fire or overheat detector electric circuit.

(e) Wiring and other components of each fire or overheat detector system in a fire zone must be at least fire-resistant.

(f) No fire or overheat detector system component for any fire zone may pass through another fire zone, unless:

(1) It is protected against the possibility of false warnings resulting from fires in zones through which it passes; or

(2) Each zone involved is simultaneously protected by the same detector and extinguishing system.

(g) Each fire detector system must be constructed so that when it is in the configuration for installation it will not exceed the alarm activation time approved for the detectors using the response time criteria specified in the appropriate TSO or an acceptable equivalent, for the detector.

Part II Analysis, Study and Exercises

Words

accumulation	[əˌkjuːmjəˈleɪʃn]	*n*. 积累;堆积
adjacent	[əˈdʒeɪs(ə)nt]	*adj*. 邻近的;毗连的;接近的
adverse	[ˈædvəːs]	*adj*. 不利的;相反的;敌对的
alternatively	[ɔːlˈtɜːnətɪvlɪ]	*adv*. (引出第二种选择或可能的建议)要不,或者
arcing	[ˈɑːkɪŋ]	*n*. 电弧
auxiliary	[ɔːgˈzɪlɪərɪ]	*adj*. 辅助的;(发动机、设备等)备用的
clogging	[ˈklɒgɪŋ]	*v*. (使)阻塞,堵塞;塞满(湿厚物);阻碍;(血或乳脂)凝结成块; 跳木鞋踢踏舞(clog 的现在分词) *adj*. 阻塞了的
closure	[ˈkləʊʒə]	*n*. 关闭;终止,结束
compartment	[kəmˈpɑːtmənt]	*n*. [建] 隔间;区划;分隔间,隔层
compliance	[kəmˈplaɪəns]	*n*. 顺从,服从;符合;屈从;可塑性; 遵守
contamination	[kənˌtæmɪˈneɪʃən]	*n*. 污染;污染物
creep	[kriːp]	*v*. 匍匐;爬行; 蔓生;蔓延
dependent	[dɪˈpendənt]	*adj*. 依靠的;依赖的; 受……的影响;取决于
detection	[dɪˈtekʃ(ə)n]	*n*. 侦查;察觉;发觉;检查
deterioration	[dɪˌtɪərɪəˈreɪʃən]	*n*. 恶化,变坏;退化;堕落
discharge	[dɪsˈtʃɑːdʒ]	*v*./*n*. 排出(物); 放出(物)
dispatch	[dɪˈspætʃ]	*v*./*n*. 派遣,发送
disposal	[dɪˈspəʊzl]	*v*. 去掉;清除;处理; (企业、财产等的)变卖,让与
drain	[dreɪn]	*v*. 流空,排放 ;(使)流走,流出; 使(精力、金钱等)耗尽 *n*. 下水道;排水管
drainage	[ˈdreɪnɪdʒ]	*n*. 排水系统
duct	[dʌkt]	*n*. (传送液体、气体、电线、电话线等的)管道,管子; (人体或植物体内输送液体的)管,导管

exceed	[ɪk'siːd]	*vt*. 超过;胜过
fireproof	['faɪəpruːf]	*n*. 防火材料 *adj*. 防火的
firewall	['faɪəwɔːl]	*n*. 防火墙
flame-out	['fleɪmˌaut]	*n*. [航空](发动机)熄火
flammable	['flæməbl]	*adj*. 易燃的; 可燃的
fume	[fjuːm]	*n*. (浓烈的或有害的)烟,气,汽
fuselage	['fjuːsəlaːʒ]	*n*. [航空]机身
hazard	['hæzəd]	*n*. 危险,风险
hazardous	['hæzədəs]	*adj*. 危险的; 有害的
ignition	[ɪg'nɪʃn]	*n*. 点燃;点火
impingement	[ɪm'pɪndʒmənt]	*n*. 侵犯,冲击;反跳; 影响; 碰撞,冲(打、撞)击
inadvertently	[ˌɪnəd'vɜːtəntlɪ]	*adv*. 不注意地;疏忽地;无意地; 并非故意地
ingestion	[ɪn'dʒestʃən]	*n*. 吸收;咽下;摄取
inlet	[ɪn'let]	*n*. 入口,进口
insulate	['ɪnsjʊleɪt]	*vt*. 使隔离,使孤立;使绝缘,使隔热
integral	['ɪntɪgrəl]	*adj*. 必不可少的
integrity	[ɪn'tegrətɪ]	*n*. 完整;正直
interruption	[ˌɪntə'rʌpʃn]	*n*. 阻断物;中断时间; 打扰;插嘴;打岔
isolate	['aɪsəleɪt]	*vt*. 使隔离;使孤立
jeopardize	['dʒepədaɪz]	*v*. 危害,冒……的危险
leakage	['liːkɪdʒ]	*n*. 泄露
maneuver	[mə'nuːvə]	*n*. 机动;演习;策略; 调遣
moisture	['mɒɪstʃə]	*n*. 水汽
monoxide	[mə'nɒksaɪd]	*n*. [无化]一氧化物
originate	[ə'rɪdʒɪneɪt]	*v*. 起源,引起
overflow	[ˌəʊvə'fləʊ]	*v*. 漫出;溢出; 挤满了人; 扩展出界;过度延伸 *n*. 容纳不下的人(或物);溢出的液体; 溢出,上溢(运算产生的数值位数或字的长度等超过存储单元的长度)
overhaul	[ˌəʊvə'hɔːl]	*v*./*n*. (机器的)解体检修;(系统、体制等的)全面修改
parameter	[pə'ræmɪtə]	*n*. 参数;范围
plenum	['pliːnəm]	*n*. 充满;充满物质的空间
propulsion	[prə'pʌlʃn]	*n*. 推进;推进力

provision	[prəˈvɪʒn]	*n*. 规定;条款;提供;准备
pyrotechnic	[ˌpaɪrəʊˈteknɪk]	*adj*. 烟火的;令人眼花缭乱的;出色的
quantity	[ˈkwɒntətɪ]	*n*. 量;数量;大量
remainder	[rɪˈmeɪndə]	*n*. 剩余物
rotate	[rəʊˈteɪt]	*v*. 旋转;循环
rupture	[ˈrʌptʃə]	*v*./*n*. 破裂; 裂开
soak	[səʊk]	*n*. 浸;湿透
sparking	[ˈspɑːkɪŋ]	*n*. 发火花;点火 *v*. 导致(spark 的现在分词);发出火星;使大感兴趣
spillage	[ˈspɪlɪdʒ]	*n*. 溢出
stabilize	[ˈsteɪbəlaɪz]	*v*. 使安定;使坚固
stall	[stɔːl]	*n*. [航空]失速
strike	[straɪk]	*v*. 撞击;打;踢(球等);突击;罢工;使打动;侵袭;突然想到;照射;击出(火星);擦(火柴);触发(电弧);鸣;弹奏;开采出;行进;铸造(硬币);(电影摄影术)复刻;达到(平衡);(金融)结算;闯出新的事业;撤(营);给……印象;把……迷住;摆出(姿态) *n*. 罢工;抗拒;袭击;殴打;(体育中得分的)击球;(十柱保龄球)一投全倒;(尤指石油的)发现;打击;(对上钩的鱼)急拉钓线;(投球)未击中;(投出的)好球;败坏名声的事
surge	[sɜːdʒ]	*n*. [航空]喘振
sustained	[səˈsteɪnd]	*adj*. 持续的;持久的;持久不变的 *v*. 维持;承受
tank	[tæŋk]	*n*. [航空]油箱
unattended	[ˌʌnəˈtendɪd]	*adj*. 无人的;无人照看的
vapor	[ˈveɪpə]	*n*. 蒸气;潮气;雾气
ventilate	[ˈventɪleɪt]	*v*. 使(房间、建筑物等)通风;使通气
ventilation	[ˌventɪˈleɪʃən]	*n*. 通风系统, 通风,通气
vent	[vent]	*n*. (空气、气体、液体的)出口,进口,漏孔 *v*. 表达,发泄(感情,尤指愤怒)
vibration	[vaɪˈbreɪʃn]	*n*. 颤动,震动,振荡, 共鸣
withstand	[wɪθˈstænd]	*v*. 承受;抵住;顶住;经受住

Phrases & Expressions

adjacent part	相邻部分；邻近部分
air intake system	进气系统
air intake system duct	进气系统管道
airflow delivery	引气输送
atmospheric temperature	大气温度
bleed air duct	排气管道
carbon monoxide contamination	一氧化碳污染
constant attention	持续关注
cooling test	冷却试验
correction factor	修正因数；修正系数
critical condition	临界条件；临界状态；临界运行条件
design precaution	设计预防措施
detection method	检测方法
deviate from	偏离;脱离
discharge line	排放管道
draw air from	从……进气/引气
essential APU's	基本的 APU
exceed established limits	超出既定限制
excessive deflection	过度的变形
excessive internal pressure	内部压力过大；内压过高
exhaust gas	废气；排出的气体
exposure to	曝光;显露
extinguishing agent	灭火剂
extinguishing agent container	灭火瓶
fire hazard	火灾危险；着火危险
fire resistant	耐火的
fire spreading to other regions	火灾蔓延到其他区域
fireproof shield	防火的屏蔽件
flammable fluid	易燃液体；燃油
fuel flowmeter	燃油流量计
fuel leakage	燃油泄漏
ground attitude	地面运行姿态
hazardous quantity	危险量
heat and corrosion resistant	耐热和耐腐蚀的

ignite flammable fluids	点燃可燃液体
ignition of leaking flammable fluid	点燃漏出的可燃液体
impingement of the fluids or vapors	液体或蒸汽的渗入
in compliance with	符合,服从
inability to operate	无法操作
inertia loads	惯性载荷
inflight start capability	在空中启动能力;飞行起动能力
ingestion of foreign matter	吸入外来物
manual valves	手控阀
maximum continuous power	最大连续功率
minimize the hazard	最小化风险;危害减至最小
non-essential APU's	备用 APU 件
non-fireproof part	非防火零件
personnel compartment	人员舱室;载人舱
pressure relief	减压;降压;压力释放;释压
prevent bursting	防止爆裂
reasonable assurance	合理保证
reverse flow	逆流;回流
rotating shaft	旋转轴;回转轴
safeguard against	预防;防止
sea level conditions	海平面状况;海平面条件
starting, stopping, and emergency shutdown	起动、停车和应急关断
subject to	使臣服;使顺从;遭受;受……影响
supply the air	供给空气
sustained operation	持续运行;持续工作
unattended operation	无人操作;无人值守
unintentional operation	无意操作;误操作
unusable fuel	未能消耗的剩余燃油;不可用油量

Abbreviations

APU	Auxiliary Power Unit	辅助动力装置
TSO	Technical Standard Order	技术标准规定

Sentence Comprehension

1. An essential APU is defined as an APU whose function is required for the dispatch of the airplane and/or continued safe flight.

   译文: 必要的 APU 是指为了保证飞机的调度或持续安全飞行所必需的 APU。

2. A non-essential APU is defined as an APU whose function is a matter of convenience, either on the ground or in flight, and may be shut down without jeopardizing safe airplane operation.

译文：非必要的 APU 是指能够在地面或飞行中提供便捷，同时可以在不危及飞机安全运行的情况下关闭的 APU。

3. The APU system must be designed and installed to give reasonable assurance that those APU operating limitations that adversely affect rotor structural integrity will not be exceeded in service.

译文：APU 系统的设计和安装必须能合理保证，在服役中不会超过对 APU 转子结构完整性有不利影响的 APU 使用限制。

4. Each fuel system must be constructed and arranged to ensure a flow of fuel at a rate and pressure established for proper APU functioning under each likely operating condition, including any maneuver for which certification is requested and during which the APU is permitted to be in operation.

译文：每一种燃料系统的构造和布置都必须能够确保燃料的流速和压力符合 APU 在每一种可能的操作条件下正常运行的要求，包括要求认证和允许 APU 运行的任何操作。

5. The quantity of fuel in the tank may not exceed the amount established as the unusable fuel supply for that tank under the requirements of § 25.959 plus that necessary to show compliance with this section.

译文：油箱内的燃油量不得超过第 25.959 条规定的该油箱不可用油量与验证本条符合性所需的油量之和。

6. Each fuel line must be installed and supported to prevent excessive vibration and to withstand loads due to fuel pressure and accelerated flight conditions.

译文：每条燃油管路的安装和支承必须能防止过度的振动，并能承受燃油压力及加速度飞行所引起的载荷。

7. The air intake system for the APU must supply the air required by the APU under each operating condition for which certification is requested.

译文：辅助动力装置的进气系统必须在合格审定要求的每种运行条件下均能提供辅助动力装置所需的空气。

8. Each air intake system duct must be drained to prevent accumulation of hazardous quantities of flammable fluid and moisture in the ground attitude.

译文：每个进气系统的管道必须排空，以防止在地面姿态下积聚危险量的易燃液体和湿气。

9. Each exhaust system must ensure safe disposal of exhaust gases without fire hazard or carbon monoxide contamination in any personnel compartment.

译文：必须确保每个排气系统对废气的安全处置，在任何人员舱内都不得存在火灾隐患或一氧化碳污染。

10. Piping must be supported to withstand any vibration and inertia loads to which it would be subjected in operation.

译文：排气管必须有支承，以承受运行中会遇到的任何振动和惯性载荷。

11. Each control must be able to maintain any set position without creep due to control loads, vibration, or other external forces resulting from the location.

译文：操纵器件必须能保持在任何给定的位置而不会由于操纵载荷、振动或其他的外力而滑移。

12. Electrical equipment subject to arcing or sparking must be installed to minimize the probability of contact with any flammable fluids or vapors that might be present in a free state.

译文：易产生电弧或火花的电气设备的安装必须能最大限度地减少与游离状态下易燃液体或蒸气接触的可能性。

13. There must be complete drainage of each part of each designated fire zone to minimize the hazards resulting from failure or malfunctioning of any component containing flammable fluids.

译文：指定火区的每个部位必须能完全排放积存的油液，以最大限度地减少任何含有易燃液体的组件的故障或故障造成的危险。

14. There must be a means for each fire extinguishing agent container to indicate that the container has discharged or that the charging pressure is below the established minimum necessary for proper functioning.

译文：对于每个灭火瓶都必须设有指示措施，指示该灭火瓶已经释压或其充气压力低于正常运行时所需的最小规定值。

15. Design precautions must be taken to minimize the hazards to the airplane in the event of an APU rotor failure or of a fire originating within the APU which burns through the APU casing.

译文：在 APU 转子故障或源自 APU 内部的火灾烧穿 APU 机壳时，必须采取设计预防措施，以使对飞机的危害降到最低。

Advanced Word Study

accessory [əkˈsesərɪ] *n*. 附件；配件；附属物；(衣服的)配饰；从犯；同谋
adj. 辅助的；副的

Definitions:

① an extra piece of equipment that is useful but not essential or that can be added to something else as a decoration

② a thing that you can wear or carry that matches your clothes, for example a belt or a bag

③ not the most important when compared to others

Synonyms: (*adj.*) auxiliary, supplemental, supplementary

(*n.*) add-on, adjunct, appliance, attachment, option

Antonyms: chief, main, principal

Usage examples:

① There is a shop with bicycle *accessories.*

② The supermarket offered furnishings and *accessories* for the home.

③ APU-mounted *accessories* must be approved for installation on the APU concerned and use the provisions of the APU for mounting.

auxiliary [ɔːgˈzɪlɪərɪ] *adj.* 辅助的;(发动机、设备等)备用的

Definitions:

① a worker who gives help or support to the main group of workers

② of a piece of equipment used if there is a problem with the main piece of equipment

Synonyms: additional, assistant, helping, attached, deputy, spare

Usage examples:

① *Auxiliary* workers/services may be provided upon request.

② The system shall contain an *auxiliary* pump.

③ The air induction system for each engine and auxiliary power unit must supply the air required by that engine and *auxiliary* power unit under each operating condition for which certification is requested.

compartment [kəmˈpaːtmənt] *n.* [建] 隔间;区划;分隔间,隔层

Definitions:

① one of the separate sections which a coach/car on a train is divided into

② one of the separate sections that something such as a piece of furniture or equipment has for keeping things in

Synonyms: chamber, room, space, coach

Usage examples:

① I opened the door of the *compartment* in the train.

② The desk has a secret *compartment.*

③ There is a handy storage *compartment* beneath the oven.

④ He found an empty first-class *compartment.*

⑤ Fireproof within the APU *compartment* and for a sufficient distance upstream of the APU compartment to prevent hot gas reverse flow from burning through the APU air intake system ducts and entering any other compartment or area of the airplane in which a hazard would be created resulting from the entry of hot gases.

contamination [kənˌtæmɪˈneɪʃn] *n.* 污染;污染物

Definitions:

① the process or fact of making a substance or place dirty or no longer pure by adding a substance that is dangerous or carries disease

② (formal) the act of influencing people's ideas or attitudes in a bad way

Synonyms: pollution, empoison, impurity

Usage examples:

① We are all worried about radioactive/ bacterial/chemical *contamination.*

② Environmental *contamination* resulting from human activities is a major health concern.

③ They feared cultural *contamination* from the influx of tourists.

④ The *contamination* of the sea around the island may be just the beginning.

⑤ Each exhaust system must ensure safe disposal of exhaust gases without fire hazard or carbon monoxide *contamination* in any personnel compartment.

critical [ˈkrɪtɪkl] *adj.* 批评的;批判性的;挑剔的;极重要的;关键的;严重的;不稳定的;可能有危险的

Definitions:

① expressing disapproval of somebody/something and saying what you think is bad about them

② extremely important because a future situation will be affected by it

③ serious, uncertain and possibly dangerous

④ involving making fair, careful judgements about the good and bad qualities of somebody/something

⑤ according to the judgement of critics of art, music, literature, etc.

Synonyms: safe, uncritical

Antonyms: vital, serious, judgemental

Usage examples:

① The newspaper often publishes *critical* comments and reports.

② The supervisor is always very *critical.*

③ It was considered to be a *critical* factor in the election campaign.

④ Reducing levels of carbon dioxide in the atmosphere is of *critical* importance.

⑤ One of the victims of the fire remains in a *critical* condition.

⑥ Students are encouraged to develop *critical* thinking instead of accepting opinions without questioning them.

⑦ It must be established over the range of operating conditions for which certification is required, that the APU installation vibratory conditions do not exceed the *critical* frequencies and amplitudes established under the TSO.

⑧ The cooling tests must be conducted with the airplane in the configuration, and operating under the conditions that are *critical* relative to cooling.

discharge [dɪsˈtʃɑːdʒ] *n*. 排出(物); 放出(物) *v*. 释放;排出;解雇;放出

Definitions:

① to give somebody official permission to leave a place or job

② to make somebody leave a job

③ to release force or power

④ to do everything that is necessary to perform and complete a particular duty

⑤ to fire a gun, etc.

⑥ the action of releasing a substance such as a liquid or gas

⑦ a substance that comes out from inside somewhere

Synonyms: blasting, fire, shot, emission, loose, bleeding

Antonyms: inhalation

Usage examples:

① The new leader has already *discharged* his duties/responsibilities/obligations.

② Patients were being *discharged* from the hospital too early.

③ Lightning is caused by clouds *discharging* electricity.

④ The factory was fined for *discharging* chemicals into the river.

⑤ Arrangements have been made for the *discharge* of payments.

⑥ Each ventilation means must be arranged so that no *discharged* vapors will cause an additional fire hazard.

⑦ Unless the extinguishing agent capacity and rate of *discharge* are based on maximum air flow through a zone, there must be means to allow the crew to shutoff sources of forced ventilation to any fire zone.

impinge [ɪmˈpɪndʒ] *v*. 对……有明显作用(或影响);妨碍;侵犯

Definitions:

① to strike or dash especially with a sharp collision

② to have an effect: make an impression

Synonyms: bang, bump, collide, crash, hit, impact, knock, smash, strike, swipe

Usage examples:

① I heard the rain *impinged* upon the earth.

② Waiting for a new idea to *impinge* upon my mind.

③ *Impinge* on other people's rights.

jeopardize [ˈdʒepədaɪz] *v*. 危害,冒……的危险

Definitions: to risk harming or destroying something/somebody

Synonyms: endanger, harm, compromise

Antonyms: safeguard

Usage examples:

① He would never do anything to *jeopardize* his career.

② This scandal could seriously *jeopardize* his chances of being re-elected.

③ A non-essential APU is defined as an APU whose function is a matter of convenience, either on the ground or in flight, and may be shut down without *jeopardizing* safe airplane operation.

④ Where continued rotation of an APU could *jeopardize* the safety of the airplane, there must be a means for stopping rotation.

prompt [prɒmpt] *adj*. 立即;迅速的;及时的;敏捷的;准时的

v. 促使;导致;激起;鼓励,提示,提醒(某人说话);给(演员)提词

n. (给演员的)提词,提示;提示符

adv. 准时地

Definitions:

① to move to action

② being ready and quick to act as occasion demands

③ performed readily or immediately

④ something that reminds about anything or something

Synonyms: quick, ready, apt

Antonyms: restrain, discourage

Usage examples:

① Evidence *prompting* an investigation.

② *Prompt* assistance is very welcomed.

③ *Prompt* emergency medical care was very important in his case.

④ There must be approved, quick acting fire or overheat detectors in each APU compartment in numbers and locations ensuring *prompt* detection of fire.

sustain [səˈsteɪn] *v*. 维持;承受

Definitions:

① to provide enough of what somebody/something needs in order to live or exist

② to make something continue for some time without becoming less

③ to experience something bad

④ to provide evidence to support an opinion, a theory, etc.

⑤ to support a weight without breaking or falling

Synonyms: maintain, bear, support, uphold

Usage examples:

① I only had a little chocolate to *sustain* me on my walk.

② When she lost her job they could no longer *sustain* their expensive lifestyle.

③ She managed to *sustain* everyone's interest until the end of her speech.

④ The company *sustained* losses of millions of dollars.

⑤ The ice will not *sustain* your weight.

⑥ Each fuel system for an essential APU must be capable of *sustained* operation throughout its flow and pressure range likely to be encountered in operation.

Exercises for Self-Study

Ⅰ. Translate Sentences

1. The components of the installation must be constructed, arranged, and installed so as to ensure their continued safe operation between normal inspections or overhauls.
2. If the tests are conducted under conditions deviating from the maximum ambient atmospheric temperature, the recorded APU temperatures must be corrected under paragraph (c) of this section.
3. The cooling tests must be conducted with the airplane in the configuration, and operating under the conditions that are critical relative to cooling.
4. Each exhaust system component must be ventilated to prevent points of excessively high temperature.
5. Be designed and located so that it cannot be inadvertently operated by persons entering, leaving, or moving normally in the flight deck.
6. Each control must be located so that it cannot be inadvertently operated by persons entering, leaving, or moving normally in the area of the control.
7. Each flammable fluid shutoff means and control must be fireproof or must be located and protected so that any fire in a fire zone will not affect its operation.
8. Each compartment must be constructed and supported so that it can resist any vibration, inertia, and air load to which it may be subjected in operation.
9. No fire or overheat detector may be affected by any oil, water, other fluids, or fumes that might be present.
10. Each component that hot exhaust gases could strike, or that could be subjected to high temperatures from exhaust system parts, must be fireproof.

Ⅱ. Multiple Choice Questions

1. The components of the installation must be ________, ________, and ________ so as to ensure their continued safe operation between normal inspections or

overhauls.

A. arranged B. constructed C. implemented D. installed

2. Each fuel system for an essential APU must be capable of ________ operation throughout its flow and pressure range.

A. continuous B. sustained C. intermittent D. extended

3. Each exhaust shroud must be ventilated or ________ to avoid a temperature high enough to ignite any flammable fluids or vapors external to the shroud.

A. replaced B. uninstalled C. insulated D. isolated

4. Each designated fire zone must be ventilated to prevent the ________ of flammable vapors.

A. accumulation B. increase C. decrease D. creation

5. Each component of the stopping system located in the APU compartment must be at least ________.

A. hermetic B. water resistant C. precise D. fire resistant

6. The ________ of fuel in the tank may not exceed the amount established as the ________ supply for that tank under the requirements of § 25. 959 plus that necessary to show ________ this section.

A. quality B. quantity C. unusable fuel
D. correction factor E. compliance with

7. Each exhaust system must ensure safe disposal of exhaust gas without fire hazard or ________ in any personnel compartment.

A. carbon dioxide contamination
B. nitric oxide accumulation
C. nitrogen dioxide contamination
D. carbon monoxide contamination

8. Each fire detector system must be constructed so that when it is in the configuration for installation it will not exceed that alarm activation time ________ for the detector using the response time criteria specified in the appropriate ________ or an acceptable ________, for the detector.

A. approved B. approving C. approves
D. TSO E. TC F. TPP
G. equality H. equivalence I. equivalent

Ⅲ. Fill in the Blanks

1. Bleed air duct failure between the APU and the airplane unit served by the bleed air, unless it can be shown that no ________ exists to the airplane.

2. The installation must comply with the installation ________ provided under the TSO.
3. The components of the installation must be constructed, arranged, and installed so as to ________ their continued safe operation.
4. It shall be possible to shut down the APU from the flight deck in normal and ________ conditions.
5. Each component of the stopping system located in the APU compartment must be at least ________.
6. It must be established over the range of operating conditions, that the APU installation vibratory conditions do not ________ the critical parameters.
7. The test may be performed in flight or on the ground under closely ________ flight conditions.
8. Each fuel line must be installed and supported to prevent excessive vibration and to ________ loads due to fuel pressure and accelerated flight conditions.
9. Flexible hose must be approved or must be shown to be ________ for the particular application.
10. A maximum ambient atmospheric temperature corresponding to sea ________ conditions must be established.
11. The materials used to form the remainder of the air intake system duct and plenum chamber of the APU must be capable of resisting the ________ heat conditions likely to occur.
12. No exhaust ________ may discharge so as to cause a fire hazard with respect to any flammable fluid vent or drain.
13. The portion of each APU control located in a designated ________ zone must be at least fire resistant.
14. The closing of any fuel shutoff valve for any APU may not make fuel ________ to the main engines.
15. The discharge end of each discharge line must be located so that discharge of the fire ________ agent would not damage the airplane.
16. There must be means to allow the crew to check the ________ of each fire or overheat detector electric circuit.

Ⅳ. Grammar/Logical Mistakes Correction

1. If the tests are conducted under conditions deviated from the maximum ambient atmospheric temperature, the recorded APU temperatures must be corrected.
2. The cooling tests must be conducted with the airplane in the configuration, and

operating under the conditions that are critical relative with cooling.

3. Each duct connected with components between which relative motion could exist must have a means for flexibility.
4. Except as providing in paragraph (b), each line, fitting, and other component carrying flammable fluid in any area subject to APU fire conditions, and each component which conveys or contains flammable fluid in a designated fire zone must be fire resistant, except for that flammable fluid tanks and supports in a designated fire zone must be fireproof.
5. Each shutoff valve must have a mean to relieve excessive pressure accumulation unless a mean for pressure relief is otherwise provided in the system.
6. There must be approved, quick acting fire or overheat detectors in each APU compartment in locations assuring prompt detection of fire.
7. Design precautions must be taken to minimize the hazards to the airplane in the event of an APU rotor failure or of a fire originating within the APU to burn through the APU casing.

Ⅴ. True or False

1. Operation of any shutoff that may interfere with the later emergency operation of other equipment is acceptable.
2. Some specific gases may discharge so as to cause a fire hazard with respect to any flammable fluid vent or drain.
3. Each fire zone involved is spontaneously protected by the same fire detector and fire extinguishing system.
4. The APU air inlet system is accepted to cause vibration harmful to the APU because of the air-flow distortion.
5. It doesn't matter whether to shut down a non-essential APU during the flight or keep it working on.
6. The temperature lapse rate is 3.6 ℃ per thousand feet of altitude above sea level until a temperature of −69.7 ℃ is reached, above which altitude, the temperature is considered constant at −69.7 ℃.

Ⅵ. Questions to Answer

1. How shall major components of the installation be particularly bonded to other parts of an airplane?
2. What are the requirements for the fuel system if an APU can be supplied with fuel from more than one tank?

3. What specifies the end of cooling tests for each critical condition?
4. What conditions may challenge the safety of the exhaust piping?
5. Under what circumstances should all components, including ducts, within a designated fire zone be fireproof?

Ⅶ. Questions for Discussion

1. What are the functions of the APU?
2. Which are the conditions that we need to provide means to automatically shutdown the APU?
3. What should be paid attention to when the APU working under hot weather?
4. Can an aircraft be operated with a malfunctioning APU? If so, any necessary restrictions needed?
5. Should the APU have the same lifespan as engines? Why?

Appendix A Definitions and Abbreviations (定义和缩略语)

[from DOC9760 Airworthiness Manual Part 1]

A.1 Definitions(定义)

Aeronautical product: Any aircraft, aircraft engine, aircraft propeller or a part to be installed thereon.

航空产品:任何航空器、航空器发动机、航空器螺旋桨或安装在其上的部件。

Aeroplane: A power-driven heavier-than-air aircraft, deriving its lift in flight chiefly from aerodynamic reactions on surfaces which remain fixed under given conditions of flight.

飞机:由动力驱动的重于空气的航空器,它在飞行中的升力来源主要是在给定的条件下空气动力学对固定表面的反作用。

Aeroplane system: An aeroplane system includes all elements of equipment necessary for the control and performance of a particular major function. It includes both the equipment specifically provided for the function in question and other basic related aeroplane equipment such as that required to supply power for the equipment operation. The engine is not considered to be an aeroplane system.

飞机系统:飞机系统包括为控制和实施特定的主要功能所需的设备的全部因素。它包括为所述功能特别提供的设备和与飞机有关的其他基本设备,如为设备的运行要求而提供的动力。发动机并不被认为是飞机系统。

Air Operator Certificate (AOC): A certificate authorizing an operator to carry out specified commercial air transport operations.

航空运营人许可证(AOC):授权运营人执行特定的商业航空运输业务的许可证。

Aircraft: Any machine that can derive support in the atmosphere from the reactions of the air other than the reactions of the air against the earth's surface.

航空器:能在大气中从空气的反作用而不是空气对地球表面的反作用中得到支持的任何机器。

Aircraft operating manual: A manual, acceptable to the State of the Operator, containing normal, abnormal and emergency procedures, checklists, limitations, performance information, details of the aircraft systems and other material relevant to the operation of the aircraft.

(Note: The aircraft operating manual is part of the operations manual.)
航空器使用手册:运营人所在国可以接受的手册,它包含正常、非正常与应急程序、检查单、限制、性能资料、航空器系统的详细内容以及与航空器运行有关的其他材料。(注:航空器使用手册是运行手册的一部分。)

Airworthiness Directive (AD): A regulatory document which identifies aeronautical products in which an unsafe condition exists, and where the condition is likely to exist or develop in other aeronautical products of the same type design. It prescribes mandatory corrective actions to be taken or the conditions or limitations under which the aeronautical products may continue to be operated. The AD is the common form of mandatory continuing airworthiness information mentioned in Annex 8.
适航指令(AD):指明存在不安全状况的航空产品的监管文件,而且这种状况在同一设计型号的其他航空产品中也存在或可能会发展。它规定采取强制性纠正措施或对航空产品继续运行提出条件或限制。适航指令是附件 8 中提到的强制性持续适航信息的通常形式。

Airworthiness Standards: Detailed and comprehensive design and safety criteria applicable to the category of the aeronautical product (aircraft, engine and propeller) that satisfy, at a minimum, the applicable standards of Annex 8.
适航性标准:适用于航空产品的类别(航空器、发动机和螺旋桨)至少能满足适用的附件 8 的详细的和全面的设计及安全标准。

Airworthy: The status of an aircraft, engine, propeller or part when it conforms to its approved design and is in a condition for safe operation.
适航:符合批准设计和安全运行条件的航空器、发动机、螺旋桨或部件的状态。

Appropriate airworthiness requirements: The comprehensive and detailed airworthiness codes established, adopted or accepted by a Contracting State for the class of aircraft, engine or propeller under consideration.
相应适航要求:缔约国为航空器、发动机或螺旋桨的等级所确定、通过或接受的全面而详细的适航规范。

Certificate holder: An individual or organization that meets the established requirements and functions at the level of competency and safety required by the State to undertake an aviation-related activity for which it has been licensed, certified, authorized and/or approved to perform.
证书持有者:在从事经许可、认证、授权和/或批准的与航空有关的活动并在国家要求的能力和安全水平方面,达到指定的要求和功能的个人或组织。

Certification basis: The applicable airworthiness and environmental standards established by a State as the basis by which the type design of an aeronautical product, or change to that type design, is approved or accepted. The certification basis may also include special conditions of airworthiness, findings of equivalent level of safety, and/or exemptions when determined by the State to apply to the type design.

合格审定依据:国家建立的适用的适航性和环境标准,作为批准或接受航空产品的型号设计或对该型号设计的改动的依据。合格审定依据还包括适航性、发现的相等的安全水平和/或国家使用这一型号设计时决定的豁免。

Certification maintenance requirement: Scheduled maintenance that is required by design to help show compliance with the appropriate type certification basis by detecting the presence of a safety-significant latent failure that would result in a hazardous or catastrophic failure condition.

合格审定维修要求:设计中应包括定期的维修计划,以帮助通过检测由危险或灾难性的故障条件造成的潜在的重大安全故障,从而表明遵守适当的型号合格审定。

Configuration Deviation List (CDL): A list established by the organization responsible for the type design with the approval of the State of Design which identifies any external parts of an aircraft type which may be missing at the commencement of a flight, and which contains, where necessary, any information on associated operating limitations and performance correction.

构型偏离清单(CDL):由负责型号设计的机构编制的、经设计国批准的清单,它指明在飞行开始时某型号的航空器上可以缺少哪些外部部件,并在必要时包含相关使用限制和性能修订的任何资料。

Continuing airworthiness: The set of processes by which an aircraft, engine, propeller or part complies with the applicable airworthiness requirements and remains in a condition for safe operation throughout its operating life.

持续适航性:航空器、发动机、螺旋桨或零件为符合适用的适航要求并在整个使用寿命期间保持安全运行状态所遵守的整套流程。

Environmental Standards: The specifications defined in Annex 16—Environmental Protection for the certification of aircraft noise and engine smoke and gaseous emissions, including the standards for the prevention of intentional fuel venting into the atmosphere.

环境标准:附件16——针对合格审定航空器的噪声和发动机烟雾以及气体排放制定的环境保护标准,包括为阻止故意把燃料排放到大气中而制定的规定。

Equivalent level of safety: As used in type certification, a finding where literal compliance with a specific airworthiness requirement cannot be demonstrated but compensating factors exist in the type design that can be shown to provide a level of safety equivalent to that intended by the certification basis.

等效的安全水平：用于型号合格审定，不能证明在文字上直接符合特定适航要求，但在型号设计中存在补偿因素，它可以提供合格审定依据所期望的同等安全水平。

Exception/Exemption：A relief from compliance with the requirement (s) of airworthiness or environmental standards, or operating rules, based on the determination by a civil aviation authority that granting such relief will not adversely affect safety.

例外/豁免：根据认为授予这种宽恕不会影响民航局的安全的决定，它是对遵守适航性要求、环境标准或运行规则的一种宽慰。

Extended Diversion Time Operation (EDTO)：Any operation by an aeroplane with two or more turbine engines where the diversion time to an en-route alternate aerodrome is greater than the threshold time established by the State of the Operator.

延长改航时间运行(EDTO)：双发或多发涡轮飞机飞往航路备降机场的改航时间超过运营人所在国规定的阈值时间的任何运行。

EDTO —Configuration，Maintenance and Procedures (CMP) requirements：The particular aeroplane configuration minimum requirements including any special inspection，hardware life limits，Master Minimum Equipment List (MMEL) constraints and maintenance practices found necessary to establish the suitability of an airframe-engine combination for extended diversion time operation.

EDTO——构型、维修和程序(CMP)的要求：特定的飞机构型的最低要求包括任何特殊检查、硬件寿命的限制、最低主设备清单(MMEL)约束和维修措施发现有必要为延长的改航时间运行建立机身-发动机组合的适用性。

EDTO——significant system：An aeroplane system whose failure or degradation could adversely affect the safety particular to an EDTO flight，or whose continued functioning is specifically important to the safe flight and landing of an aeroplane during an EDTO diversion.

EDTO——重要的系统：飞机系统故障或退化可能影响 ETDO 飞行的安全，或其继续运转对于 EDTO 改航中的安全运行和着陆是特别重要的。

Failure condition：The effect on the aircraft and its occupants，both direct and consequential，caused or contributed to by one or more failures，considering relevant adverse operational or environmental conditions.

故障条件：考虑到相关的不良操作或环境条件，由一个或多个故障对航空器及其乘客造成的直接和间接的影响。

Instructions for Continuing Airworthiness (ICA): A set of descriptive data, maintenance planning and accomplishment instructions, developed by a design approval holder in accordance with the certification basis for the aeronautical product. The ICAs provide air operators with the necessary information to develop their own maintenance programme and also for approved maintenance organizations to establish the accomplishment instructions.

持续适航指令(ICA):由设计批准书持有人在航空产品合格审定的基础上制定的一套描述性的数据、维修大纲和产品说明。持续适航指令(ICA)除了向航空运营人提供必要的信息使其能制定自己的维修大纲外,还为批准的维修机构确立完成任务的指示。

Latent failure: A failure that is not detected and/or enunciated when it occurs.

潜在故障:在故障发生时未检测到的和/或未确定的故障。

Life-limited part: Any part for which a mandatory replacement limit (in hours, cycles or calendar time) is specified in the type design, the mandatory continuing airworthiness information or instructions for continuing airworthiness. These parts must be permanently removed from service on or before this limit is reached.

寿命有限制的零部件:在型号设计、强制性持续适航信息或持续适航指令中有强制性更换时间限制(小时、周期或日历时间)的任何零部件。在到达限制时间或之前,这些零部件必须永久换掉。

Maintenance: The performance of tasks required to ensure the continuing airworthiness of an aircraft, including any one or combination of overhaul, inspection, replacement, defect rectification, and the embodiment of a modification or a repair.

维修:为确保航空器持续适航所需执行的任务包括大修、检查、换件、纠正缺陷或其组合,以及具体的改装或修理。

Maintenance organization's procedures manual: A document which details the maintenance organization's structure and management responsibilities, scope of work, description of facilities, maintenance procedures, and quality assurance, or inspection systems. This document is normally endorsed by the head of the maintenance organization.

维修机构的程序手册:经维修机构的负责人核准的文件,其中详细载明了维修机构的结构和管理职责、工作范围、设施介绍、维修程序及质量保证制度或质量检查制度。

Maintenance programme: A document which describes the specific scheduled maintenance tasks and their frequency of completion and related procedures, such as a reliability programme, necessary for the safe operation of those aircraft to which it applies.

维修大纲:描述具体的定期维修任务及其完成的频次和相关程序(如可靠性方案)的文件,这些程序对适用它的那些航空器的安全运行是必要的。

Maintenance release: A document which contains a certification confirming that the maintenance work to which it relates has been completed in a satisfactory manner, either in accordance with the approved data and the procedures described in the maintenance organization's procedures manual or under an equivalent system.

维修放行单:证明已按照批准的数据和维修机构的程序手册所述的程序或根据与之相当的制度令人满意地完成有关维修工作的文件。

Major modification: In respect of an aeronautical product for which a type certificate has been issued, a change in the type design that has an appreciable effect, or other than a negligible effect, on the mass and balance limits, structural strength, engine operation, flight characteristics, reliability, operational characteristics, or other characteristics or qualities affecting the airworthiness or environmental characteristics of an aeronautical product.

重大改装:对已经颁发型号合格证书的航空产品的设计型号的修改会对重量和平衡的限制、结构强度、发动机的运行、飞行特点、可靠性、运行特点、其他影响航空产品的适航性或环境特点的特征或性质都有很大的影响,否则的话这种影响可以忽略不计。

Major repair: Any repair of an aeronautical product that might appreciably affect the structural strength, performance, engine, operation flight characteristics or other qualities affecting airworthiness or environmental characteristics.

大修:对航空产品进行的可能会明显影响结构强度、性能、发动机、运行飞行特点,或影响适航性或环境特点质量的任何修改。

Mandatory Continuing Airworthiness Information (MCAI): The mandatory requirements for the modification, replacement of parts, or inspection of aircraft and amendment of operating limitations and procedures for the safe operation of the aircraft. Among such information is that issued by Contracting States in the form of airworthiness directives.

强制性持续适航信息(MCAI):为了航空器的安全运行,强制要求修改、更换零件或检查航空器和航空器安全运行的限制和程序的修订。其中还有缔约国以适航性指示的形式发表的信息。

Master Minimum Equipment List (MMEL): A list established for a particular aircraft type by the organization responsible for the type design with the approval of the State of Design containing items, one or more of which is permitted to be unserviceable at the commencement of a flight. The MMEL may be associated with special operating conditions, limitations or procedures.

最低主设备清单(MMEL):由负责型号设计的机构编制并经设计国批准,为某一特定型号的航空器建立的设备清单,在这些设备中的一项或几项不能使用的情况下,航空器仍然可以开始飞行。最低主设备清单可与特定的运行条件、限制或程序相关联。

Minimum Equipment List (MEL): A list which provides for the operation of aircraft, subject to specified conditions, with particular equipment inoperative, prepared by an operator in conformity with, or more restrictive than, the MMEL established for the aircraft type.

最低设备清单(MEL):由运营人为某型号航空器编制的、在特定条件下允许某些设备不工作时实施运行的清单。该清单符合为该型航空器建立的最低主设备清单的要求,或更为严格。

Minor modification: A modification other than a major modification.

次要改装:大修以外的主要修改。

Minor repair: A repair other than a major repair.

小修:大修以外的修理。

Modification: A change to the type design of an aeronautical product which is not a repair.

改装:对航空产品型号设计的改动而不是修理。

Operations manual: A manual containing procedures, instructions and guidance for use by operational personnel in the execution of their duties.

运行手册:运行人员在履行其职责时所用的包含程序、说明和指南的手册。

Operator's maintenance control manual: A document which describes the operator's procedures necessary to ensure that all scheduled and unscheduled maintenance is performed on the operator's aircraft on time and in a controlled and satisfactory manner.

运营人维修管理手册:描述运营人的程序的文件,这些程序对于确保及时地、有管理地和令人满意地完成对运营人的航空器的所有定期和不定期维修是必要的。

Organization responsible for the type design: The organization which is the holder of the type certificate and has the responsibility of the design of the aeronautical product and the continuous compliance of the aeronautical product type design to the appropriate airworthiness requirements imposed by the type certificating authority. In some cases, it will be the holder of an equivalent document certifying approval of the type design by the certificating authority.

负责型号设计的机构:持有型号合格证书的机构,负责航空产品的设计以及航空产品的型号设计持续符合型号合格审定当局提出的有关适航要求。在某些情况下(附件

8 第 98 次修订之前)，它是证明型号设计获得合格审定当局批准的同等文件的持有人。

Propulsion system: A system consisting of an engine and all other equipment utilized to provide those functions necessary to sustain, monitor and control the power/thrust output of any one engine following installation on the airframe

推进系统:包含发动机和所有其他设备的一个系统，安装在机身上之后用于提供必要的功能，以维持、监测和控制任何一个发动机的动力/推力。

Repair: The restoration of an aeronautical product to an airworthy condition as defined by the appropriate airworthiness requirements.

修理:将航空产品恢复到符合有关适航性要求所界定的适航条件。

Safety management system: A systematic approach to managing safety, including the necessary organizational structures, accountabilities, policies and procedures.

安全管理体系:管理安全的系统做法，包括必要的组织结构、问责制、政策和程序。

State of Design: The State having jurisdiction over the organization responsible for the type design.

设计国:对负责型号设计的机构拥有管辖权的国家。

State of Manufacture: The State having jurisdiction over the organization responsible for the final assembly of the aircraft.

制造国:对负责航空器最后组装的机构拥有管辖权的国家。

State of Registry: The State on whose register the aircraft is entered.

登记国:航空器在其登记册上登记的国家。

State of the Operator: The State in which the operator's principal place of business is located or, if there is no such place of business, the operator's permanent residence.

运营人所在国:在该国有运营人的主要营业场所，如果没有这样的业务经营场所，则为运营人的永久居住地。

Structural inspection: A detailed inspection of the airframe structure that may require special inspection techniques to determine the continuous integrity of the airframe and its related parts.

结构检查:对机身结构的详细检查，它可能要求用特殊的检验技术来确定机身及其有关部件的持续的完整性。

Threshold time: The range, expressed in time, established by the State of the Operator to an en-route alternate aerodrome, whereby any time beyond requires an EDTO approval from the State of the Operator.

阈值时间:运营人所在国规定的飞往航路备降机场的航程(以时间表示),超过这一航程的任何时间均需要得到运营人所在国对延长改航时间运行的批准。

Type certificate: A document issued by a Contracting State to define the design of an aircraft type and to certify that this design meets the appropriate airworthiness requirements of that State.
(Note: Some Contracting States also issue Type Certificates for engines and propellers.)
型号合格证:缔约国为确定某一航空器型号设计,并证明该设计满足该国相应适航要求所颁发的文件。
(注:一些缔约国还为发动机和螺旋桨颁发型号合格证。)

Type design: The set of data and information necessary to define an aeronautical product type for the purpose of airworthiness determination to any later aeronautical product of the same type.
型号设计:确定航空产品型号所需要的一套数据和信息,目的是为同一型号以后的航空产品确定适航性。

A.2　Abbreviations(缩略语)

AD	Airworthiness Directive	适航指令
AED	Airworthiness Engineering Division	适航工程部门
AFM	Aircraft Flight Manual	航空器飞行手册
AID	Airworthiness Inspection Division	适航检查部门
ALI	Airworthiness Limitation Items	适航性限制项目
AMO	Approved Maintenance Organization	经批准的维修机构
AOC	Air Operator Certificate	航空运营人许可证
APU	Auxiliary Power Unit	辅助动力装置
CAA	Civil Aviation Authority	民用航空当局
CoA	Certificate of Airworthiness	适航证
CoR	Certificate of Registration	登记证
CDL	Configuration Deviation List	构型偏离清单
CG	Centre of Gravity	重心
CMR	Certification Maintenance Requirements	合格审定维修要求
DGCA	Director General of Civil Aviation	民航局局长
EDTO	Extended Diversion Time Operations	延长改航时间运行
ETOPS	Extended-Range Operations (by Aeroplanes with Two Turbine Engines)	(双发)延程运行
ICA	Instructions for Continuing Airworthiness	持续适航指令

Kg	Kilogram	千克
LoV	Limit of Validity	有效性的限制
MCAI	Mandatory Continuing Airworthiness Information	强制性持续适航信息
MCM	Maintenance Control Manual	维修管理手册
MEL	Minimum Equipment List	最低设备清单
MMEL	Master Minimum Equipment List	最低主设备清单
MOPM	Maintenance Organization's Procedures Manual	维修机构程序手册
MRB	Maintenance Review Board	维修审查委员会
MSG	Maintenance Steering Group	维修指导小组
MTOM	Maximum Certificated Take-Off Mass	最大审定起飞质量
OEM	Original Equipment Manufacturer	原始设备制造商
RVSM	Reduced Vertical Separation Minimal	减少的最低垂直间隔
SB	Service Bulletin	服务公告
SDR	Service Difficulty Report	服务困难报告
STC	Supplemental Type Certificate	补充型号合格证
TBO	Time Between Overhauls	大修间隔时间
TC	Type Certificate	型号合格证
TCB	Type Certification Board	型号合格审定委员会
TSN	Time Since New	自从新启用以来
TSO	Technical Standard Order	技术标准规范

Appendix B　Answers to Exercises for Self-Study

Chapter 1

Ⅰ. Translate Sentences

1. 航空器应满足附件 8 的适航性要求，但如果不满足附件 6 的附加要求，则可能不能用于特定的运行任务。
2. 基本航空立法应该包含建立民航当局主动监督和监管民用航空活动的规定。
3. 采用另一个国家的规定有一些优点，但只能在确保其他国家的规定与国际民航组织的标准相一致以后才能予以考虑。
4. 新的检查员应与有经验的检查员搭档，以确保在职培训被执行并载入文件。
5. 公约的第 31 条规定，适航证书由飞机注册地所在国签发或使之有效。
6. 有了成文的法规，一国可以准予该法规的例外或豁免，前提是该国的标准中存在确定该例外或豁免的机制。
7. 如果授予了豁免，那么应当附带条件和限制，包括豁免的时间限制。
8. 人们还认识到，一个国家或多国集团可以选择通过与区域安全监督组织或机构达成协议来履行其职责。
9. 至关重要的是，协议应明确界定各当事方应履行的职能，以确保所有义务都能被充分履行。
10. 这有助于适航人员确定设计国批准的强制性修改、检查和修理是否得到适当执行。

Ⅱ. Multiple Choice Questions

1. B　2. B, C　3. C　4. D　5. B　6. A, D, F
7. A　8. B　9. C　10. B　11. A, B　12. A

Ⅲ. Fill in the Blanks

1. framework　2. to enhance　3. legislation　4. supervises
5. authorization　6. validation　7. commensurate　8. accredited
9. towards　10. impending danger　11. in regard to

Ⅳ. Grammar/Logical Mistakes Correction

1. *in* which
2. *are equal to or above* the minimum standards
3. and *to* carry out (the action is still acted by the CAA)
4. a mechanism *within* the States criteria
5. including a *time* limitation

6. *so* as to ensure
7. *in* behalf of
8. and *keep* pace with
9. to keep files *for*

Ⅴ. True or False

1. F 2. T 3. T 4. T 5. T 6. F 7. T 8. F

Ⅵ. Questions to Answer

1. Annex 6—Operations of Aircraft, Annex 8—Airworthiness of Aircraft, Annex 16—Environmental protection.
2. The State.
3. AED and AID.
4. Nomination.
5. A new inspector should be evaluated on successfully performing tasks in accordance with CAA requirements.
6. All air operators and to CAAs located in States that have the affected aeronautical product on their respective national aircraft register.
7. AID.

Chapter 2

Ⅰ. Translate Sentences

1. 飞机的配置可能随着重量、高度和温度的变化而变化,在一定程度上,它们符合本章节(f)所要求的操作程序。
2. 在所制定的该飞机使用操作限制范围内,起飞数据必须包括下列操作修正因素。
3. 飞机的重量、构型、功率(推力)在每一分段内必须保持不变,而且必须与该分段内主要的最临界的状态相对应。
4. 飞行航迹必须基于无地面效应的飞机性能。
5. 对于航路配置,本节第(b)款和第(c)款所规定的飞行路径必须在已为飞机确定的重量、高度和环境温度运行限制范围内。
6. 对于陆地飞机和水陆两用机来说,着陆距离必须在平坦、平整、干燥、坚硬的跑道上确定。
7. 即将资源的消耗量分为若干档次,不同档次按不同的标准收费,档次越高,收费标准越高。
8. 多引擎固定翼飞机的关键引擎是其故障对飞机操纵和性能造成最不利影响的引擎。
9. 当飞机的高度达到它的翼展时,可以认为飞机脱离了地面效应。
10. 这一机构承认,与其达成一致一点都算不上标准操作程序,且与我们的政策相左,我们本应当拒绝那些条款上的任务的。

Ⅱ. Multiple Choice Questions

1. A，D 2. E，A 3. B 4. C，B 5. C

Ⅲ. Fill in the Blanks

1. prescribed 2. compatible 3. prevailing 4. compensating
5. exceptional 6. equivalent 7. alertness 8. humidity
9. configurations 10. crews 11. segments

Ⅳ. Grammar/Logical Mistakes Correction

1. consistent *with*
2. *within* the operational limits
3. may *not*
4. *on* water
5. *whose* failure

Ⅴ. True or False

1. F 2. T 3. F 4. T 5. F

Ⅵ. Questions to Answer

1. The relative humidity must vary linearly.
2. No.
3. Five stages：takeoff，climb，en-route/cruise flight，descent，and landing.
4. An airplane is considered to be out of the ground effect when it reaches a height equal to its wing span.
5. If the takeoff distance does not include a clearway，the takeoff run is equal to the takeoff distance.
6. The takeoff flight path begins at the end of the takeoff distance 35 feet above the takeoff surface.

Chapter 3

Ⅰ. Translate Sentences

1. 强度和变形要求达到限制载荷和极限载荷的验证通常需要测试子组件、全尺寸组件，或完成组装组件（如接近完整的机身）的全尺寸测试。
2. 如果测试结果与分析结果不一致，应找出原因并采取适当的措施。
3. 对于每一种特定的发动机设计，申请人应该考虑这些类型的故障是否适用，以及它们是否出现比叶片飞出更严重的载荷情况。
4. 飞机设计载荷应针对滑行、起飞和着陆过程中出现的最临界情况而设计。
5. 在制定建议的检查方案时，还应考虑损伤位置和损伤扩展数据。
6. 对完整结构或主要结构部分进行测试的性质和范围将取决于先前适用的设计和结构测试，以及类似结构的服役经验。
7. 在证明熟悉组件载荷路径和潜在故障模式条件下，可利用在相同标准和方法条件下设计的类似在役组件的服役和测试方法。

8. 建立一致化的生产，有可能减少机构可接受的非关键铸件规则要求的非目视检查的次数。

9. 如果从这些试验中得到的力学性能分布是可接受的，那么与鉴定程序中确定的性能值相比，试验的频率可能会降低。

10. 面板应承受最不利的压力载荷组合，包括最大内部压力、外部空气动力压力、温度效应，以及适当的飞行载荷。

Ⅱ. Multiple Choice Questions

1. B　2. A, B, C　3. C　4. A　5. B　6. B
7. C　8. B　9. A　10. A　11. C

Ⅲ. Fill in the Blanks

1. intensities, distributions　2. previous　3. compliance　4. justifications
5. fatigue damage　6. corrosion　7. specimens　8. pitch

Ⅳ. Grammar/Logical Mistakes Correction

1. are *of* such a low order
2. *accounted* for, *supported* by
3. *extent* of tests, in *connection* with
4. *requiring* test
5. *ultimate* control

Ⅴ. True or False

1. F　2. F　3. F　4. F　5. F

Ⅵ. Questions to Answer

1. Acceptable Means of Compliance.
2. The sizing of an airplane structure, determination of flight load intensities and distributions, flight load measurements, ensuring accuracy of measurements.
3. To obtain flight loads which can represent real loads under operating conditions.
4. Flight Load Validation ensures that the results we obtained under Flight Load Measurement are acceptable.
5. Only when validated by full-scale tests.
6. When no reliable analytical methods exist.
7. Multiple load path construction and the use of crack stoppers.
8. Particularly, corrosion.
9. The effect of variability.

Chapter 4

Ⅰ. Translate Sentences

1. 燃料必须以发动机类型证书中规定的压力输送到每个发动机。

2. 油箱内的燃油量不得超过第 25.959 条规定的该油箱不可用油量与验证本条符合性所需的油量之和。
3. 飞机的重量必须是油箱满油、带有最小机组以及配重(保持重心在允许范围内所需)时的重量。
4. 必须用试验表明,装机后的油箱能承受本节第一款第(1)或(2)条所规定的压力(取大者),而不被损坏或漏油。
5. 如果通过实际发动机机油消耗量的数据证实,则可以使用高于本节第 2 款第(1)和第(2)条规定的燃油/机油比。
6. 软滑油箱必须经过批准,或必须表明适合其特定用途。
7. 每个机油散热器风道的位置必须确保在发生火灾时,来自发动机机舱正常开口的火焰不会直接冲击散热器。
8. 进行冷却试验时,飞机的形态和运行条件均必须取每一飞行阶段中对于冷却是临界的情况。
9. 将飞机部分结构空间(例如,机翼)加以密封用以存贮燃料的油箱。

Ⅱ. Multiple Choice Questions

1. B 2. A, B, C 3. B 4. B 5. C 6. B
7. C 8. D, B 9. A, E 10. A 11. C 12. B

Ⅲ. Fill in the Blanks

1. improbable 2. requirements 3. type 4. approved
5. difference 6. specimen 7. compartment 8. isolated
9. independent 10. failure

Ⅳ. Grammar/Logical Mistakes Correction

1. *necessary* for, for *which* compliance, substituted *for*
2. *operating* experience
3. must be *vented*
4. *to strike*
5. free *of*
6. resulting *from*

Ⅴ. True or False

1. F 2. F 3. T 4. T

Ⅵ. Questions to Answer

1. Fuel tank vents and the fuel transfer system shall be designed.
2. From the top part.
3. A drain.

Chapter 5

Ⅰ. Translate Sentences

1. 申请航空器型号合格证时，必须附有该航空器的三视图和现有的初步基本数据。
2. 如果申请人选择遵守在提交型号合格证申请后生效的对本节的修订，申请人也必须遵守局方认为与之直接相关的任何其他修订。
3. 本节第23、27、31、33和35部分中所包含的适航要求，或局方认可的其他适航标准，适合并适用于特定设计和预期用途，并提供局方可接受的安全水平。
4. 申请人可以将特殊检查和预防性维护计划作为航空器型号设计的一部分或补充型号设计。
5. 任何其他必要数据将被允许使用，以便通过比较来确定同类型后续产品的适航性、噪声特性、燃料排气和废气排放(如适用)。
6. 每个申请人必须允许适航当局进行任何必要的检查，以及任何必要的飞行和地面测试，以确定是否符合《联邦航空条例》的适用要求。
7. 对于装有以前未在认证航空器中使用的涡轮发动机的飞机，必须对全套发动机进行至少300小时型号合格审定飞行测试。
8. 申请正常类、实用类、特技类、通勤类或运输类飞机类型证书的申请人必须提供一名持有合适飞行员证书的人，以进行本部分所要求的飞行测试。
9. 每个申请人必须向局方提交一份符合性声明(FAA表格317)，以证明提交给局方进行型号认证的每架航空器的发动机和螺旋桨都符合规定。
10. 型号设计包括图纸和说明书，以及这些图纸和说明书的清单，这些图纸和说明书是定义所示产品的配置和设计特征所必需的，以符合本子章中适用于该产品的那部分要求。

Ⅱ. Multiple Choice Questions

1. B 2. A 3. A, B, C 4. B 5. D, A

Ⅲ. Fill in the Blanks

1. maintenance 2. acceptance 3. technical 4. instructions
5. accomplished 6. accuracy 7. include

Ⅳ. Grammar/Logical Mistakes Correction

1. must be accompanied *by*
2. necessary *to show*, product *to be certificated*
3. *surrendered*, *suspended*, *revoked*
4. compliance *with*
5. or *supplemental* type design

Ⅴ. True or False

1. F 2. F 3. F 4. T 5. T

Ⅵ. Questions to Answer

1. The drawings and specifications; information on dimensions, materials, and processes; Airworthiness Limitations of the Instructions for Continued Airworthiness; a special inspection and preventative maintenance program; determination of the airworthiness, noise characteristics, fuel venting, and exhaust emissions.
2. The type design, inspection reports, and computations.
3. Special conditions.

Chapter 6

Ⅰ. Translate Sentences

1. 适航证书只有在向局方提出申请后才可以修改。
2. 航空器的所有人、经营人或受托人应根据要求将航空器提供给局方检查。
3. 按照本章第43.15款规定的100小时性能检查规则对航空器(之前根据本节颁发了一种不同适航证的实验性验证航空器除外)进行了检查,符合适航要求。
4. 局方确定该航空器符合经批准的型号设计并处于安全运行状态。
5. 对于目前可能不符合适航要求但能够安全飞行的航空器,可签发特别飞行许可证。
6. 持证者根据第135部获权操作他们所运营的飞机,并根据第135.411款第a(2)或(b)条规定的持续适航维护程序维护飞机。
7. 将飞机飞往须进行修理、改装或保养的基地,或飞行至贮存处。
8. 本条款下的许可证可发给根据本章第121部授权的证书持有人。
9. 审定司可进行或要求申请人进行适当的检查或必要的测试。
10. 尽管有本部分所有的其他规定,但标准适航证书的初始核发必须遵守下列规定。

Ⅱ. Multiple Choice Questions

1. C　2. A, B, C　3. A, C, D　4. C　5. B　6. A, D　7. A, B, C
8. C　9. B

Ⅲ. Fill in the Blanks

1. amended　2. type　3. effective　4. conforms to
5. evidence　6. authorized　7. certificate

Ⅳ. Grammar/Logical Mistakes Correction

1. *came* into effect
2. aircraft *having*, that *meets*
3. *in* which
4. *is limited to* the additional fuel
5. conforms *to*
6. entitled *to*

Ⅴ. True or False

1. T　2. T　3. F　4. F　5. T

Ⅵ. Questions to Answer

1. Despite, in spite of.
2. Decibel, a unit used to measure the intensity of a sound.
3. When an aircraft may not currently meet applicable airworthiness requirements but is capable of safe flight.
4. Federal Aviation Administration.
5. Only one.

Chapter 7

Ⅰ. Translate Sentences

1. 安装的各部件的构造、布置和安装必须保证在正常检修或大修的间隔内能继续保持安全运转。
2. 如果在偏离最高外界大气温度的条件下进行试验，则必须按本条(c)修正所记录的 APU 温度。
3. 进行冷却试验时，飞机的形态和运行条件均必须取每一飞行阶段中对于冷却是临界的情况。
4. 所有排气系统部件均必须通风，以防某些部位温度过高。
5. 操纵器件的位置必须保证不会由于人员进出驾驶舱或在驾驶舱内正常活动而使其误动。
6. 每个控制装置必须被安装在确保在控制装置区域内人员进入、离开和正常移动都不会使控制器被无意运转的位置。
7. 可燃液体的切断装置和控制装置必须是防火的，或者必须进行定位防护，以保证火灾区域内的任何火灾都不会影响其运行。
8. 每个隔室的构造和支撑必须使其能够抵抗运行中可能受到的任何振动、惯性和气动载荷的作用。
9. 火警或过热探测器不得受到任何可能存在的油、水、其他液体或气体的影响。
10. 可能受到热废气冲击或可能受到排气系统部件高温影响的每个部件都必须是防火的。

Ⅱ. Multiple Choice Questions

1. B, A, D　2. B　3. C　4. A　5. D　6. B, C, E
7. D　8. A, D, I

Ⅲ. Fill in the Blanks

1. hazard	2. instructions	3. ensure	4. emergency
5. fire resistant	6. exceed	7. simulated	8. withstand
9. suitable	10. level	11. maximum	12. gases

13. fire 14. unavailable 15. extinguishing 16. functioning

Ⅳ. Grammar/Logical Mistakes Correction

1. *deviating* from
2. relative *to*
3. connected *to*
4. *provided* in, except ~~*for*~~ that
5. a *means*, a *means*
6. *ensuring* prompt
7. *which burns* through

Ⅴ. True or False

1. F 2. F 3. F 4. F 5. T 6. F

Ⅵ. Questions to Answer

1. Electrically bonded.
2. Appropriate manual switching capability is needed to prevent interruption of fuel flow to that APU.
3. Cooling tests can be stopped when the component and APU fluid temperatures stabilize; the stage of flight is completed; or an operating limitation is reached.
4. Heat, corrosion, failure due to expansion, vibration, inertia loads.
5. When exposed to or damaged by fire.

References

[1] ICAO Airworthiness Manual: Doc 9760 [S]. 3rd ed. ICAO,2014:27-60.

[2] 国际民用航空组织.航空器适航性手册:Doc 9760 [S].3 版.国际民用航空组织,2014: 27-56.

[3] FAA FAR Part 25—Airworthiness Standards: Transport Category Airplanes [S]. FAA,2004:Subpart B, Subpart E.

[4] FAA Appendix K to Federal Aviation Regulations FAR Part 25—Auxiliary Power Unit Installations [S]. FAA,2004:K25.901-K25.1203.

[5] CCAR-25-R4 中国民用航空规章:第 25 部 运输类飞机适航标准[S].中国民用航空局,2011: 3-13.

[6] EASA Certification Specifications and Acceptable Means of Compliance for Large Aeroplanes [S]. EASA, 2015:388-507.

[7] FAA FAR Part 21—Certification Procedures for Products and Parts [S]. 2004: Subpart B, Subpart H.

[8] 中国民用航空规章:第 21 部 民用航空产品和零部件合格审定规定[N].中国民用航空局公报, 2012:8-15.